Lecture Notes in Chemistry 68

Edited by:

Prof. Dr. Gaston Berthier
Université de Paris

Prof. Dr. Hanns Fischer
Universität Zürich

Prof. Dr. Kenichi Fukui
Kyoto University

Prof. Dr. George G. Hall
University of Nottingham

Prof. Dr. Jürgen Hinze
Universität Bielefeld

Prof. Dr. Joshua Jortner
Tel-Aviv University

Prof. Dr. Werner Kutzelnigg
Universität Bochum

Prof. Dr. Klaus Ruedenberg
Iowa State University

Prof Dr. Jacopo Tomasi
Università di Pisa

W0044176

Springer

Berlin
Heidelberg
New York
Barcelona
Budapest
Hong Kong
London
Milan
Paris
Singapore
Tokyo

A. Julg

From Atoms and Molecules to the Cosmos

A Quasi-Ergodic Interpretation
of Quantum Mechanics

 Springer

Author

Prof. André Julg
15, Avenue Massenet
F-13009 Marseille, France

Library of Congress Cataloging-in-Publication Data

Julg, André, 1926-
    From atoms and molecules to the cosmos : a quasi-ergodic
  interpretation of quantum mechanics / A. Julg.
      p.   cm. -- (Lecture notes in chemistry ; 68)
    Includes bibliographical references.
    ISBN 3-540-64636-1 (softcover : alk. paper)
    1. Quantum theory.   I. Title.   II. Series.
  QC174.12.J85   1998
  530.12--dc21   98-27932
                                                                CIP

ISBN-13: 978-3-540-64636-5   e-ISBN-13: 978-3-642-48939-6
DOI: 10.1007/978-3-642-48939-6

The use of general descriptive names, registered names, trademarks, etc. in this
publication does not imply, even in the absence of a specific statement, that such
names are exempt from the relevant protective laws and regulations and therefore
free for general use.

Typesetting: Camera ready by author
SPIN: 10553990   51/3143 - 543210 - Printed on acid-free paper

Preface

What has always struck me, by reading textbooks and publications intended for teaching as well as for research, is the inconscience or the resignation with which most of scientists of the present day do admit the coexistence of two physics, namely the classical physics and the quantum one, as if we were in the presence of two worlds completely disjoined and relevant to two entirely different interpretations.

The fact to say, as it is implicitly often admitted, that quantum mechanics applies to atoms and molecules whereas classical physics is reserved to the macroscopic scale, not only is not satisfying for the mind because, whatever its size may be, matter is built up of the same constituents, hence it is relevant to the same explanation, but also is inconsistent with experiment. Do we not know macroscopic systems which exhibit a quantum behavior, as does the liquid helium ? Are crystals not giant molecules ?

And what is worse, according to the sake of the cause, it occurs that the physicist, for the same object, utilizes one or the other of these two theories. The case of the electron is particularly typical in this regard. One reads, indeed, in the classical textbooks that any charged particle, moving with a nonuniform velocity, radiates energy. The radiation theory of the ionized belt of Jupiter and that of pulsars are precisely based on this phenomenon, likewise for the radiation theory of the synchroton. Now the power radiated by one electron describing a circular orbit of radius R with the velocity v , is proportional to $v^4 R^{-2}$. Consequently, the smaller the radius, the more the electron radiates. Still, everyone claims that in an atom the electrons do not radiate or, more precisely, that there exists states of well-determined energy for which electrons do not radiate, the fine lines of the absorption and emission spectra being interpreted as arising from jumps between these states. Now, given the size of the system, the radiation should have been very intense. Consequently, if the conventional interpretation is correct, we have the right to wonder what is the critical size from which the system does not radiate and, on the other hand, what is the cause of the discontinuity in its behavior.

The attitude of the chemist is just as ambiguous. He is working at the macroscopic level but he argues at the microscopic one, considering atoms and molecules, as well as electrons, according to the cases, for convenience sake, as being classical or quantal objects. For instance, the frequences of the vibrations of molecules and crystals are calculated from a classical model which likens the system to a set of coil-springs, whereas quantum theory is used to determine the energy levels and the intensities. Likewise, the chemical reactivity is explained, either classically from electron pairs localized between the nuclei, or quantically by considering the electrons as being completely delocalized over the whole molecule.

On the analogy of the relativity theory which gives the Newton mechanics again when the velocity of the particle is negligible with respect to that of light, it is quite possible, of course, to think that when the Planck constant, which is an ingredient peculiar to quantum mechanics, tends to zero, one obtains the classical physics. In fact, the reality is completely different. The $h \to 0$ limit does not give the classical mechanics (To give an example of difficulties encountered by this passage at the limit, we can quote the case of the kinetic terms which are vanishing if h is equal to zero) . In another way, attempts have been made in order to explain the passage from the quantum world to the classical one, h remaining constant. An interaction between the quantum system and its environment seems to be necessary (For instance, see R. Omnès : *The interpretation of Quantum Mechanics* , Princeton University Press, princeton, 1994) .

Unfortunately such interpretations essentially concern open systems constituted by one particle or non-bonded particles.

To solve the problem of the connection between the two quantum and classical worlds, it remains to investigate the inverse way which consists in trying to find the quantum formalism again from classical concepts. In this view, we will appeal to results concerning the atomic and molecular structure which is a privileged application domain of Quantum Mechanics.

Independent of the intellectual satisfaction of being able to unify the two theories, such a possibility would solve the problems which appear in understanding of phenomena considered as being quantal by nature and whose interpretation leads to a conception of the world completely opposite to that which "common sense" suggests to us. Indeed, let us remember all that has been said about the uncertainty relationships which would limit our knowledge by its fundamental indeterminism, as well as about the Bell inequality whose violation would lead to the abandonment of locality in physics. The tunnelling effect which allows a particle to be not stopped by a potential barrier whose height is greater than its energy, the electron which passes simultaneously through the two slits of the Young device, and so on, as are paradoxical situations which we should be able to clear up classically.

But, even if it is possible to explain the quantum behavior classically, the necessity to use a very peculiar formalism and not classical-like equations to be consistent with experiment, will always set a serious problem which leads us, in any case, to wonder about the physics which is hidden under this formalism.

During the last twenty years, we have published a certain number of articles concerning these problems. The aim of this book is precisely to make the synthesis of them, in view to show that, provided we accept some minor rearrangements, the quantum mechanics or rather the quantum formalism , can be effectively interpreted within the framework of the classical physics.

The structure of the book is the following. After a brief recall of quantum formalism and its main interpretations (Chap.I) , in Chapter II we propose a model based on the idea that any system is never strictly isolated, but that it is unceasingly exchanging energy with the other systems which constitute the universe. The coupling between the system under consideration and the rest of the universe entails that the various properties of the system fluctuate versus time whereas their time-average values performed over a sufficiently long time tend to finite values, which is typically the characteristic of an ergodic behavior.

A remark about the construction of the operators in Quantum Mechanics (Chap. III) shows that, contrary to what occurs in the conventional quantum formalism, the complete symmetrization of the operators with respect to the position and momentum variables (p and q , resp.) , precisely leads to fluctuations for the various properties of the system, for its energy in particular.

Chapter IV is devoted to the example of enantiomeric molecules for which the quantum description is reached, in the mean, only after a sufficiently long time. This leads (Chap. V) to the more general problem of the physical measurements which precisely give the time-average values over the duration of the measurements. Various examples are given to show that Quantum Mechanics hides the ergodic behavior of the physical systems. Chapter VI presents other examples arising from problems concerning atomic and molecular structures.

The following chapters are devoted to the mathematical exploitation of the model, by assuming (Chap. VII) that the influence of the universe over the system under consideration can be represented by an electromagnetic field constituted by uncorrelated bursts oriented in the mean along all the directions of space in an isotropic manner. By considering only the electric part of this field, a purely classical treatment allows us to find again the qualitative model presented in the preceding chapters as also the quantum operator formalism, the Schrödinger equation and the Planck constant (Chap. VIII) .

The introduction of the magnetic part of this universe field (Chap. IX) allows to find again the spin of electron and all its properties. The Stern-Gerlach experiment is explained. On the other hand, the famous Bell inequality whose violation is interpreted as showing the non-local character of Physics, is replaced by another inequality which is never violated, which restores the local character of Physics.

In Chapter X the problem of excited levels is discussed. The latter appear as being new equilibrium states of the system when it is immersed within an electromagnetic radiation. The introduction of the notion of usable energy carried by the radiation allows us to do without the photon as a physical particle.

The problem of many-particle systems is briefly discussed in Chapter XI in view of chemical applications.

The fundamental question of the wave-particle duality is discussed in Chapter XII . The de Broglie relationship is found again. But the wave associated with the particle appears as purely fictitious, only the particle possessing a physical nature.

After some consideration of the micro-irreversibility of the physical phenomena, we discuss the problem of the possible variation of the physical constants (Chap. XIV) . We show that the latter can effectively vary under the constraint that certain proportionality relationships between these constants, considered as parameters, are respected. The charge and the mass of the particles as well as the Planck constant would tend to zero with the expanding universe, so that the latter would be vanishing.

Lastly, the Conclusion dwells upon the role played by the interaction between the systems which appears as being the origin of the organization of the universe. Tackled in its whole, the universe exhibits a classical behavior, Quantum Mechanics being only a formalism which allows to take, *in the mean* , the interaction between the system under consideration and the rest of the universe into account, without it being necessary to introduce this interaction in the equations in a explicit manner.

This book, of course, is essentially aimed to all those who teach, study or utilize Quantum Mechanics. As background a minimum knowledge of this theory is recommended. Nevertheless, given the problems tackled and the conclusions which emerge from them (in particular the possibility to explain microphysics within the framework of the classical Physics and the consequences which follow from the model) , the book oversteps the limited circle of these readers. It should interest anyone who asks oneself about the very meaning of the physical theories and their implications concerning the universe.

In any case, I wish that, given the provocative (not to say revolutionary) character of the ideas developed in this book and of the conclusions obtained, this speech for the defense of a classical interpretation of Quantum Mechanics will give rise to questions and - why not ? - will form the subject of more elaborate works.

Marseille   André Julg

March 1998

Acknowledgements

I would like to thank my wife Odette for her constructive remarks and her precious help in the final writing of this book. I also thank Prof. Dr. G. Berthier (Paris) for the interest which he has taken in my work and for his comments and helpful suggestions.

CONTENTS

equation. Origin of the universality of the Schrödinger Equation. Meaning of the quantum formalism. Stabilty of atoms and molecules.

Ambiguity of the notion. The intrinsic kinetic momentum of the electron. Intrinsic magnetic momentum of electron. Magnetic momentum of positron. The Vaschy theorem. Effect of a constant magnetic field. Correlation in a singlet state. The Bell inequality.

The quantal point of view. The excited state in our model. Theorem. Consequences. The Franck-Condon principle. Relationship between the transition energy and the frequency of the radiation. Molecular spectra. Utilizable energy carried by a radiation. Induced emission and laser effect. Connexion with the perturbation theory. Remark about the states of the continuum. Thermalization effect.

Interest of the problem. The Hartree-Fock approximation. Justification of the Hartree-Fock model. Connection between the spin and the Fermi-Dirac statistics. Coming back on the orbital domains. Slater's rules . Hund's rule. Muon-electron systems.

Origin and interpretation of the concept. The spreading of a wave-packet. Wave associated with a particle. Electron diffraction. The particle in a box. Momentum associated with an electromagnetic radiation - Application to the Compton effect. Closed and unclosed systems.

The specific character of the time variable. Reversibility and irreversibility in Mechanics. Friction, irreversibility and stability. Similarity to our model. Parallelism with entropy.

The problem of the past variability of the fundamental parameters in physics. A preliminary remark. Experimental data. Choice of a unit system. Derived units. Invariance of the physical laws. Invariance of the light velocity and that of G . First consequences. Connection with the strong and weak interactions. Origin of the time-invariability of α. The principle of conservation of energy. Variation of $\hbar$ versus the expansion of the universe. Consequences and various applications. Remark about the $\hbar \to 0$ limit in quantum mechanics.

I

The quantum formalism
and its main interpretations

The axioms and their immediate consequences

As any mathematical theory, the quantum formalism is based on a certain number of axioms. Without dwelling upon the statement of the latter, we will simply recall the chief lines of the basic propositions and their most direct consequences, referring the reader to the quantum mechanics textbooks.

1° To any *state* of the system, there corresponds a function $\Psi(q,t)$ of the coordinates q of the particles which constitute the system, and of the time t . The latter, called *wave-function* , is assumed to be normalized to unit within its geometrical definition domain D

$$\int_D \Psi^* \Psi dv = 1 \tag{I-1}$$

We will remark that Ψ is defined apart from a phase factor.

2° With every time-independent dynamical property $G\,(q\,,p\,)$, there is associated a time-independent operator $\hat{G}$. Let g_i and ϕ_i be the eigenvalues and the eigenfunctions of this operator (In order to simplify we will assume that these eigenvalues are not degenerate). As for the function Ψ, we will assume that the ϕ_i's are normalized, which does not restrict the problem.

3° The result of the measurement of G is necessarily equal to one of the eigenvalues g_i of $\hat{G}$, obtained with the probability $|c_i|^2 = |\langle \Psi \phi_i \rangle|^2$.

It results from this that for a great number of measurements, the obtained average value is

$$\bar{g} = \sum_i |c_i|^2 g_i = < \Psi \hat{G} \Psi >$$ (I-2)

with $\Psi = \sum_i c_i \phi_i$.

Given that the result of a measurement is necessarily a real number, the operators $\hat{G}$ must be self-adjoint. In fact, practically, the hermiticity only is demanded, which is a weaker but sufficient condition.

Equation (I-2) can be interpreted to two different manners, according to the signification attributed to the word "system" in axiom 1. If the system is in reality a set of identical microsystems, $\bar{g}$ is the average value of the values corresponding to each microsystem at the time of the measurement. On the contrary, if the word "system" designates a well-defined microsystem, $\bar{g}$ is the average value of the values obtained by successive measurements performed on the microsytem under consideration, assuming, of course, that the latter is not perturbed by the measurement.

4° Let us now consider an isolated system. Its energy is constant. Consequently, whatever the interpretation assigned to (I-2), the measurement of the energy does necessarily give a well-determined value $\bar{E}$, which signifies that Ψ is proportional to one of the eigenfunctions of the operator $\hat{H}$ associated with energy, called Hamiltonian operator , ϕ_a, corresponding to the eigenvalue $E_a = \bar{E}$ ($|c_a| = 1$).

More generally, for an isolated system

$$\hat{H}\psi = E\psi$$ (I-3)

where ψ is a function of the coordinates only. That is the time-independent Schrödinger equation.

For such a system, Eq. (I-2) becomes

$$\bar{g} = < \psi \hat{G} \psi >$$ (I-4)

5° Formally, the operators $\hat{G}$ are deduced from the classical expression of the corresponding properties G by replacing the components of the momentum p_u ($u = x$, y, z) by the corresponding operators $\dfrac{\hbar}{i}\dfrac{\partial}{\partial u}$. For instance, for a system constituted by a single particle

$$\hat{G} = G\left(x, y, z, \frac{\hbar}{i}\frac{\partial}{\partial x}, \frac{\hbar}{i}\frac{\partial}{\partial y}, \frac{\hbar}{i}\frac{\partial}{\partial z}\right) \tag{I-5}$$

In fact, as we will see in Chapter III, this rule is equivocal. Nevertheless, for a first approach of the quantum formalism, we will be content with this rule.

Rule (I-5) shows, in particular, that the operator associated with a property depending on the coordinates x, y, z only, exhibits the same expression as the property itself. For instance

$$op(x^n) = x^n \tag{I-6}$$

A particularly important notion is that of the probability density, i.e. the average probability to find the particle at a given time in the vicinity of a given point of the space. In order to obtain the average value of x^n, namely $\overline{x^n} = <\Psi x^n \Psi>$, it is necessary that the probability density be equal to

$$\rho = |\Psi|^2 \tag{I-7}$$

which gives a physical meaning to the square of the modulus of the wave-function.

The time-invariance of the various properties of an isolated system involves that

$$\Psi(q,t) = \psi(q)\exp\left[if(t)\right] \tag{I-8}$$

In particular, for the density (I-7)

$$|\Psi|^2 = |\psi|^2 \tag{I-9}$$

From this brief outline of the principles on which the quantum formalism is based, the chief point which emerges is that the information contained in the wave-function is sufficient to obtain the value of the various properties of the system without requiring the detailed knowledge of the motion of the particles, as in classical mechanics. In particular, the notion of trajectory does not appear, as well as the instantaneous value of a given property in the case where the latter would vary versus time.

Such a situation raises the problem of the physical meaning of this formalism and, consequently, that of the interpretation of quantum mechanics.

The various interpretations

Although the theorists as well as the experimentalists are unanimous in acknowledging the power of quantum mechanics and its operational validity, no consensus has been possible up to present time concerning its interpretation and the underlying physics. Sometimes, the discussion turned into a veritable polemic, for instance between Bohr and Einstein in the thirties. At the present time, even, a certain dogmatism is not bad taste.

Roughly, the problem focalizes around two strongly tangled questions:

i. Does quantum mechanics apply to a well-defined unique system, or to a set of identical systems ?

i.i. Is quantum mechanics complete ? In other words, is the information which it allows to obtain, the sole one which is accessible to us owing to the nature of the things or does there exist more detailed information concerning the behavior of the particles which quantum mechanics cannot reach owing to some limitation inherent in its structure ?

Concerning the first question, we have previously remarked that the axioms of quantum mechanics do not make the exact nature of the system precise, and that the average value $\bar{g}$ of a property (I-2) can be interpreted according to two different ways, either by considering a single system, or a set of identical systems. In fact, the second question is bound with the first one because, if quantum mechanics applies to a set of systems considered as a whole, it is obvious that, as in any statistical theory, quantum mechanics cannot afford a fine description of the behavior of each of the particles of the system, so that it is *incomplete* , which would involve the existence of *hidden variables* which would vanish in the global treatment, although they play an essential role in the physics of the system.

In fact, although the statistical interpretation keeps its supporters [1] , it raises a certain number of difficulties. First, what do signify the variables q ? It is difficult to understand how a mechanics which would only introduce the coordinates of one microsystem, would be able to afford results concerning the set of the microsystems without introducing any interaction between these systems. Moreover, it seems to be difficult to admit that a classical behavior of the microsystems is able to lead to a "quantal" behavior. The existence of an absorption spectrum built up upon sharp lines is typical in this regard. On the other hand, we know well-defined systems which typically exhibit quantum properties (crystals, liquid helium, magnetical materials) militates against a statistical interpretation. We can also quote recent experiments performed on a single electron [2] or a single atom (a baryum cation) .

Rejecting the statistical interpretation, the Copenhagen School (so called because its most active protagonist was Bohr [3]) accepts the quantum behavior as a whole, claiming even that everything which does not appear in quantum mechanics, does not correspond to any reality (e.g.: the notion of trajectory) . On the other hand, the passage of a particle through a potential barrier whose height is greater than the energy of the particle (tunnelling effect) is not considered as being shocking.

The conceptual difficulties which seem to appear, would be, in fact, the consequence of preconceptions which arise from our macroscopic perceptions. No reason would exist to assert that the microscopic world behaves in the same manner as the phenomena we observe on the macroscopic scale.

In this framework, quantum mechanics appears as being *complete* Any intervention of hidden variables is *a priori* excluded.

As a support of this categoric assertion - not to say of this act of faith - the von Neumann theorem [4] is put forward. According to this theorem, indeed, the existence of hidden variables is inconsistent with the mathematical structure of the quantum formalism. This point of view is, nevertheless, not universally accepted. In particular, Bohm which constructed a hidden variable model [5] , considered that the von Neumann proof is circular. For his part, Bell thought that this theorem is built up upon restrictive hypotheses which are not necessary, so that, finally, the question of the hidden variables is still delated.

Einstein, Podolsky and Rosen, in a famous article of the thirties [6] , showed the difficulties created by the Copenhagen interpretation in the form of a paradox involving a " Gedanken Experiment". A chemical illustration of the EPR paradox was given by Bohm [7] : A diatomic molecule, initially formed in a state of total spin zero from two atoms of spin one-half is " disintegrated" by a method not influencing the spin of each of them. Given that the total spin remains zero at infinity, the measurement of any component of the spin of one atom allows us to immediately conclude that this spin component is opposite in the second atom.This involves the existence of large-distance correlations, physically difficult to understand, on which we will come back in Chap. IX. Bohr rejected the paradox by stating that a quantum process cannot be analyzed, by nature, into parts located in definite regions of space and time. Nevertheless, Bell [8] showed much later that a hidden-variable theory can reproduce the quantum results only by introducing such correlations.

Whatever that may be, in spite of all the problems raised by the Copenhagen interpretation (the so-called *orthodox* interpretation) , the latter, in the past as well in the present time, has been adopted by the largest number of physicists.

From another point of view, Louis de Broglie [9] claims that the particles are *piloted* by a physical wave whose amplitude is proportional to that of the Schrödinger wave-function, except at the near vicinity of the particle where it would be very great. In order to account for the quantum behavior and, in particular, for its undetermined character, de Broglie introduces a random perturbation arising from a so-called *subquantal* medium, omnipresent in space. This medium would be the cause of unceasing jumps of the particles from one trajectory to the other, so that the probability density is proportional to the square of the modulus of the wave-function. Consequently, such a model implicitly admits that quantum mechanics is not statistical and is based on the existence of hidden variables which would be, in a certain manner, materialized by the subquantal medium, in a permanent interaction with all the physical systems. This theory, therefore, introduces a new capital element with respect to the other ones, namely the conventional character of the notion of "isolated" system. Nevertheless, the nature of the subquantal medium remains well mysterious, its properties being not clearly defined.

A very near idea has been utilized as a basis of *Stochastics Electrodynamics* (SED) [10] . In this theory, which has appeared in the sixties, the system is considered as being immersed within a universal field, the so-called *vacuum field* . The stability of the various atomic and molecular systems would arise from the balance between the energy which the systems receive from this field, and that which they radiate according to the laws of classical electrodynamics. Consequently, that is an essentially classical model. In order to insure the invariance of the vacuum field under a Lorentz transformation, the energy spectral density of this field must be proportional to ω^3. From such a distribution of energy, this essentially classical theory allows to find again all the results obtained by quantum mechanics for the harmonic oscillator (energy, spatial distribution, absorption spectrum) , and, consequently, the Planck formula for the black body. Great hopes have been set on SED by a number of authors which saw an alternative theory for quantum mechanics in this model. Unfortunately, SED was powerless to find again the quantum results for other systems, in particular for the anharmonic oscillator and the hydrogen atom [11] .

More generally, several authors wondered whether quantum mechanics could be reduced to a stochastic theory [12] . The conclusion is the following. In order to find again the so-called quantum behavior from a classical model, it is necessary, sooner or later, to introduce a more or less random element in the latter. Nevertheless, the origin of this random element has not been made precise, so that, all things considered, the problem is only shifted without being solved for all that.

At last, for memory, we will quote the many-worlds interpretation proposed by Everett [13] , which, although it is logically valid, is just a speculation without any connection to physical reality. Therefore, we will not enlarge upon this interpretation. A survey of various alternative theories on Quantum Mechanics can be found in Ref 14 .

Practical conclusion

If quantum mechanics applies effectively to a single well-defined system and not to a set of identical systems, we are in front of the serious problem which is set by the conception of the universe proposed by quantum mechanics. Have we blindly to accept the latter, as the Copenhagen School demands to us, and to resign ourselves to be ignorant of a more detailed knowledge of the universe ? That would be to lay quantum mechanics down as a true myth, with all the taboos which that involves. A myth, indeed, is only a tranquillizing explanation of the natural phenomena in view to attenuate the distressing character of a reality which is not clear for us, but which does exist nevertheless, even if we can only have a more or less complete perception of it. Science, of course, at the limit, is perhaps only a marvelous myth, but, nevertheless, we have not to exagerate for all that.

In short, to resolve the problem we must start from scratch again in a completely different mind in order to build up a unitary theory, directly from a physical model, without introducing *ad hoc* concepts whose object would be precisely to orientate the constructing of the model towards the conclusions we will obtain. The validity of the model will be judged by its possibility to find again the quantum formalism whose physical signification will become clear, and, of course, by its agreement with experiment.

References

[1] L.E. Ballentine, *Rev. of Modern Phys.* **42** (1976) 358.

[2] R.S. van Dyck, P.B. Schwinberg, H.G. Dehmelt, in *The electrons* , D. Hestness and A. Weingartshofer eds (Klower Acad. Publ., Netherlands) 1991, p. 239.

[3] N. Bohr, *Atomic Theory and the Description of Nature* (Cambridge Univ. Press) 1934 ; *Atomic Physics and Human Knowledge* (Wiley, New York) 1958.

[4] J. von Neumann, *Mathematische Grundlagen der Quanten Mechanik* (Verlag-Springer, Berlin) 1932.

[5] D. Bohm, J. Bub, *Rev. Mod. Phys.* **38** (1966) 453, **40** (1968) 235.

[6] A. Einstein, B. Podolsky, N. Rosen, *Phys. Rev.* **47** (1935) 777.

[7] D. Bohm, in *Quantum Theory* , vol. III, edited by D.R. Bates (Academic Press) 1962, p. 351.

[8] J.S. Bell, *Physics* **1** (1964) 195.

[9] L. de Broglie, in L. de Broglie, J.L. Andrade e Silva, *La réinterprétation de la Mécanique Ondulatoire* (Gauthier-Villars, Paris) 1971.

[10] for instance, see: T.W. Marshall, *Proc. Roy. Soc. London* **A, 276** (1963) 475 ; E. Santos, *Nuovo Cimento* **B, 19** (1974) 57 ; **22** (1974) 201 ; T H. Boyer, *Phys. Rev.* **D, 11** (1975) 790; P. Claverie, S. Diner, *Int. J. Quant. Chem.* **12**, suppl. **1** (1977) 41.

[11] P. Julg, *Folia Chemica Theoretica Latina* **6** (1978) 99 ; L. Pasquera, P. Claverie, *J. Math. Phys.* **23** (1982) 1315.

[12] for instance, see : P. Claverie, S. Diner, in *Localization and Delocalization in Quantum Chemistry* , O. Chalvet ed. (Reidel Publ. Co, Dordrecht) 1976, vol. III, p. 395 ; S.M. Moore, *Found. of Phys.* **9** (1979) 237 ; L. de la Peña, *ibid.* **12** (1982) 1017 ; E.B. Stechel, E.J. Heller, *Ann. Rev. Phys. Chem.* **35** (1984) 563.

[13] H. Everett, *Rev. Mod. Phys.* **29** (1957) 454 ; **42** (1976) 358.

[14] M. Jämmer, *The Philosophy of Quantum Mechanics* (J. Wiley, New York) 1074; D. Wick, *The infamous Boundary : Seven decades of Controversy in Quantum Physics* (Birkhäuser, Basel) 1995.

II

Proposal of a new model

The apparent failure of the classical physics

In order to understand the problem clearly, it is necessary to come back to the situation as it appeared at the beginning of the century. Classical physics was at the zenith with the analytical mechanics and the Maxwell equations. Rutherford had just demonstrated the atomic constitution. The notion of molecule began to become clear. All seemed to be ready to interpret the structure of matter so that the disappointment was very great when it appeared that the classical physics was not able to explain the stability of hydrogen, the simplest of the atoms, constituted by a single electron moving around a proton.

The argument was the following. Let $+e$ and $-e$ be the charges of proton and electron respectively. According to Coulomb's law, the force which is acting between the two particles is attractive and its intensity varies as the square of the distance between them

$$\mathbf{F} = -K\frac{e^2}{r^3}\mathbf{r} \tag{II-1}$$

(Subsequently, we will use the *c.g.s.* system for which the factor K is equal to 1. This factor, nevertheless, will be reintroduced in Chapter XIII).

The motion equation of the electron around the proton is formally analogous with that which governs the motion of the planets around Sun, if we neglect their interactions. Under suitable initial conditions, the electron would describe an ellipse around the proton (assumed as being infinitely heavy) which would be located at one of the foci of ellipse. Given that the initial conditions can vary continuously within a large domain, an infinity of stable electron-proton systems would exist, exhibiting different properties, which is inconsistent with experiment which shows that all the hydrogen atoms exhibit the same properties.

In order to improve the description of the system, we can introduce the damping force [1]

$$\mathbf{f} = \frac{2e^2}{3c^3} \dddot{\mathbf{r}}$$
(II-2)

which leads to the following differential equation

$$m\ddot{\mathbf{r}} = \mathbf{F} + \mathbf{f}$$
(II-3)

The consequence of the introduction of this force is that, owing to the radiated energy, the electron orbit contracts more and more so that the electron falls on the nucleus after a finite time. Here still, the conclusion is unacceptable given that the hydrogen atom is stable.

Owing to these results, it was tempting to conclude that classical electromagnetics does not apply to the infinitely small systems. As we know, this assertion is still circulating, strongly deep-rooted in the minds after a century of quanta theory. In fact, the situation is worth examining more closely before delivering such a sentence.

Origin of the stability of atoms and molecules

Instead of considering the unrealistic case of a strictly isolated atom, i.e. within an empty universe, let us rather consider the real world, constituted by atoms and molecules surrounded by other atomic and molecular systems.

Given that the hydrogen atom radiates energy, it is obvious that any system built up upon charged particles does also radiate, so that all the atoms and molecules of the universe and all the free charged particles of the latter, radiate. From which it results that the "vacuum", in fact, is filled by an electromagnetic field we will call the *universe field* .

In their motion around the nuclei, the electrons of atoms and molecules which are immersed within this field, absorb energy which balances that radiated, so that, given the unceasing   exchange of energy between the universe field and each of the various systems of the universe, an equilibrium state is reached [2] . We see, therefore, that there is no objection to the possibility for a classical model to be able to explain the stability of atoms and molecules, under the condition we abandon the too simplistic concept of *isolated system* . We have only to deal with coupled systems, the universe taken as a whole, being the unique isolated system. This explains that, if we will fictively treat a system of finite size as an isolated system, we must use a special dynamics to introduce this coupling artificially in a more or less veiled manner. It is besides interesting to recall that Slater [3] , in 1924, wrote: " Any atom may, in fact, be supposed to communicate with other atoms all the time it is in a stationary state, by means of a virtual field radiation". But this idea was not retained.

The general properties of the universe field

Owing to the fact that the universe field is created by the whole of the charges which are moving in the universe, this field appears as being a necessary physical entity. Consequently, it is conceptually very different from the field that SED introduces a priori. In this theory, not only the problem of the origin of the field do not appear, but also it is considered by certain authors, Boyer [4] for instance, as being teleological, just as the question of the existence of matter. The same remark can be also made for the zero-point field introduced in the quantum field theory for reasons which are more formal than physical.

Given its origin, the universe field we introduce must exhibit properties which have to be the reflect of the general properties of the universe. In particular, as the universe, this field must be isotropic and homogeneous on a large scale, so that its electric and magnetic components, **E** and **H** respectively, must exhibit the same average characteristics at any point of space, and vary in an isotropic manner both in direction and in amplitude around an average value equal to zero.

In fact, the average isotropy and homogeneity of the universe do not preclude local concentrations, for instance the galaxies which are strongly inhomogeneous with its stars and the planet systems which surround the latter. Our own existence results precisely from this inhomogeneity. As a consequence of these inhomogeneities in the distribution of matter, we could think that the universe field is far from being the same at any point of space. In fact, although the electric potential decreases as r^{-1} (which effectively favors the inhomogeneities close to the point under consideration) , owing to the fact that, in the mean, the number of systems at the distance r varies as r^2, the far regions bring the most important contribution to the universe field. On the other hand, given the great remoteness with respect to us of the systems located in theses regions, the fluctuations arising from the motion of theses systems are completely negligible. Likewise the fact that their distribution is discrete. This explains that the universe field exhibits the average isotropy and homogeneity of the universe itself. In the case of a system which would be very close to the system we intend to study, say at some Angströms, rather than to try to correct the universe field in order to account for the perturbation due to that system, it is better to consider the two systems as a whole, within the universe field whose uniform general properties are kept.

Given the impossibility in which we are to describe the motion of all the particles of the universe, it is, of course, impossible to explicit the variations of **E** and **H** versus time at a given point. From which, a non-previsible character which has not to be considered as being *indeterminist* owing to the fact that it arises from a classical physics.

The field introduced by SED was characterized by its spectral density $S(\omega)$, i.e. by the Fourier transform of the correlation function of the electric field $\overline{E(t).E(t+\theta)}$. Assuming that this density has to be invariant under a Lorentz transformation, an ω^3 density was obtained. Independently of the convergence difficulty for the total energy of the vacuum field (the integral $\int_0^\Omega \omega^3 d\omega$ is divergent when Ω tends to infinity) , the Lorentz invariance condition which is imposed, is not justified because the physical laws have to be invariant, but not the properties themselves. Moreover, the frequencies of the vacuum field have to be bounded, at least by that corresponding to the formation of an (e^+, e^-) pair

$$\hbar\omega_{max} = 2mc^2 \qquad\qquad\qquad (\text{II-4})$$

i.e. $v_{max} \cong 10^{20} s^{-1}$. Anyhow, the introduction of such a cut-off breaks down the Lorentz invariance.

In our model, we will not make the hypothesis concerning the distribution of the frequencies between 0 and ω_{max} (II-4) precise. We will only assume that the universe field can be simulated by very sudden and very brief bursts, oriented in an isotropic manner, varying both in modulus and in direction, and that the properties of the field exhibit the same average values at any point of the universe.

Direct consequences

In order to clarify the ideas, we will come back to the problem of the hydrogen atom. The electron which is submitted, on the one hand, to the Coulomb force and to the damping one, and, on the other, to the universe field, describes a very complex trajectory, passing near to the nucleus or moving off from the latter according to the effect of the field, intermixing and crossing as the string in a completely tangled ball. This trajectory fills the whole of space and passes in the vicinity of all the points an infinite number of time. A manner to represent this trajectory is to say that the field is acting by jerks, the electron jumping from a classical arc corresponding to Eq (III-3) and determined by the state in which the preceding burst has led it, towards another arc.

In passing, we will remark the difference between our model and SED. The latter, indeed, leads to elliptical trajectories which are regularly distorted in course of time by the effect of both the damping force and the vacuum field while in our model the trajectory is very irregular.

Moreover, during its motion the electron is alternatively accelerated and retarded in a sudden manner according to the quick variations of the universe field so that its energy does vary versus time.

The case of the hydrogen atoms, in fact, falls under the more general category of *coupled systems* , i.e. interacting with one another while keeping their individuality. Here, the atom under consideration interacts with all the other systems which constitute the universe.The theory of such systems (See Appendix and Ref 5) shows that, if the systems under consideration does not disintegrate (into proton and electron in the case of the hydrogen atom) , the values of all the properties of the interacting systems fluctuate versus time around time-average values which tend to finite limits when the time-interval over which the mean is performed, tends to infinity. More precisely, if we consider the property $G\ (t\ )$ of a given system, we have

$$\left| \frac{1}{\tau} \int_0^\tau G(t)dt - \overline{G} \right| < \varepsilon \qquad \text{when } \tau \to \infty \qquad \text{(II-5)}$$

$\overline{G}$ being the limiting time-average value of G corresponding to an infinite time-interval, and ε an arbitrarily small quantity.

Moreover, one shows [5] that it is possible to replace the limiting *time* average value of the property $G(t)$ (II-5) by the *space* -average value of G considered as a function of the position M of the particles of the system, through a *probability density* $\rho(M)$ defined at any point of the domain {D} offert to the system

$$\overline{G} = \int_D G(M)\rho(M)dM \quad \text{with} \quad \int_D \rho(M)dM = 1 \qquad \text{(II-6)}$$

The reader will find a concrete example of determination of the density in Appendix.

In a general manner, the behavior of systems for which condition (II-5) is verified, are called *ergodic* or rather *quasi-ergodic* [5] . The minimum time τ_e after which the average values of the various properties become stable, is called *ergodicity time* .

We will notice that this time depends on the accuracy ε we demand for the value of $\overline{G}$. For instance, let us consider a classical harmonic oscillator which is a particular case of ergodic system. Its kinetic energy is proportional to $\sin^2 \omega t$, ω being its frequency. Over the time intervall $(0,t)$ the average value of this quantity is equal to $\dfrac{1}{2}\left(1 - \dfrac{\sin 2\omega t}{2\omega t}\right)$, so that after n periods $(t \cong nT)$ the limiting value (1/2) is reached within the relative accuracy

$$\eta = \frac{1}{2\omega nT} = \frac{1}{4\pi n} \qquad \text{(II-7)}$$

Consequently, if we will fix the value of η , the ergodicity will be reached after $1/4\pi\eta$ periods (e.g. if $\eta = 10^{-2}$, $\tau_e = 8$ periods) .

Throughout this book, we will show how important is the ergodicity concept for the interpretation of the physical phenomena.

The example of oscillators given in Appendix shows that in the general case the motion exhibits an irregular character which very quickly becomes chaotic, i.e. unforeseeable, as soon as the number of coupled systems is sufficiently great. That is precisely the situation which occurs for the hydrogen atom which is interacting with all the other systems of the universe. So that, even if the behavior of the electron of this atom appears to us as being completely indeterminist, in fact, it is governed by the classical electrodynamics, i.e. by determinist equations.

Nevertheless, the model does not involve an absolute determinism. In particular, we remain free to act according to our will and, thereby, to modify our near environment, i.e., theoretically, to engage the universe in a new way which, in the absence of our action, it would not have followed. But, in fact, our actions will always be completely negligible with respect to the whole lot of the forces which are acting in the universe, so that the effect of our interventions will remain without consequence on the general evolution of the universe.

In any case, the strong connection which appears in our model between matter and the universe field is worth being emphasized. Without matter, indeed, this field cannot exist, and conversely, so that the question to know whether it is possible to detect the universe field is either senseless or trivial because we do exist !

References

[1] J.A. Jackson, *Classical Electrodynamics* (Wiley, New York) 2th ed., 1975, p.560.

[2] A. Julg, *La Liaison chimique* (Presses Univers. France, Paris) 1980.

[3] J.C. Slater, *Nature* **113** (1924) 307.

[4] Th. Boyer, *Phys. Rev.* **D, 11** (1975) 790.

[5] A. Blanc-Lapierre, P. Casal, A. Tortrat, *Méthodes mathématiques de la Mécanique Statistique* (Masson, Paris) 1959.

III

A remark about the construction of the operators in quantum mechanics

The operator associated with the square of the energy

Purely qualitative considerations have led us to conclude that the energy of an isolated system does fluctuate versus time, which is, of course, in complete disagreement with quantum mechanics which claims that any system exhibits states of well-defined energy, called *eigenstates*, corresponding to the eigenfunctions of the time-independent Schrödinger equation (I-3).

The proof of this proposal is based on the fact that the operator associated with the square of the energy E^2, is assumed to be equal to the square of the operator associated with the energy

$$op(E^2) = (\hat{H})^2 \qquad \text{(III-1)}$$

Thus the corresponding average quadratic dispersion is equal to zero

$$(\Delta E)^2 = \overline{E^2} - (\overline{E})^2 = \langle \psi \hat{H}^2 \psi \rangle - \langle \psi \hat{H} \psi \rangle^2 = 0$$

Relationship (III-1) is, in fact, a particular case of the general rule used by quantum mechanics

$$op(A^2) = (\hat{A})^2 \qquad \text{(III-2)}$$

Owing to the fact that, on the other hand, quantum mechanics gives a rule to construct any operator directly from the classical expression of the property under consideration, it is necessary to examine the general problem of the construction of the operators a little more closely.

The difficulties to find a general construction rule

The usually given rule to built up the operator $\hat{G}$ associated with the property G (I-5), namely that $\hat{G}$ is directly obtained from the classical expression $G(p,q)$ by replacing the components p_u $(u = x, y, z)$ of the momentum of each particle in the expression of G by the corresponding operators $\dfrac{\hbar}{i}\dfrac{\partial}{\partial x}$, ..., is far from being sufficient. On the one hand, the rule is equivocal. Indeed, terms as $x^2 p_x^2$ and $p_x^2 x^2$ are classically identical. Now, owing to the fact that x and p_x do not commute, these terms lead to different operators. Moreover, the obtained operators are not hermitian.

This difficulty did not escape to Schrödinger [1] which proposed a more or less complete symmetrization of the classical expression with respect to x and p_x before applying the rule (I-5). For instance, for xp_x^2, he proposed

$$(xp_x^2 + p_x^2 x)/2 \, , \, p_x x p_x \, , \, (xp_x^2 + p_x x p_x + p_x^2 x)/3 \qquad \text{(III-3)}$$

The corresponding operators are well hermitian, but which of them to choose ? The ambiguity remained entire.

Other authors have examined the problem. We will only give some examples:

For $x^n p_x^m$, Weyl and subsequently Moyal [2] recommend

$$2^{-n} \sum_{k=0}^{n} \binom{n}{k} x^{n-k} p_x^m x^k \qquad \text{(III-4)}$$

Margenau and Hill [3]

$$\frac{1}{2}(x^n p_x^m + p_x^m x^n) \qquad \text{(III-5)}$$

Born and Jordan [4]

$$\frac{1}{n+1} \sum_{k=0}^{n} p_x^k x^n p_x^{m-k} \qquad \text{(III-6)}$$

More recently, Boyer [5] utilized the following completely symmetrized expression

$$\frac{1}{6}(x^2 p_x^2 + x p_x^2 x + p_x^2 x^2 + x p_x x p_x + p_x x p_x x + p_x x^2 p_x) \qquad \text{(III-7)}$$

for the product $x^2 p_x^2$.

The reader will find other expressions in Ref. [6].

All these recipes lead to hermitian operators, but, excepted (III-5), the corresponding average quadratic dispersions $(\Delta E)^2$, i.e. $\overline{E^2} - (\overline{E})^2$, are different from zero. For instance, for the harmonic oscillator whose energy is of the form

$$E = \frac{1}{2m} p_x^2 + \frac{1}{2} k x^2 \qquad \text{(III-8)}$$

the square of the energy is

$$E^2 = \frac{1}{4m^2} p_x^4 + \frac{1}{4} k^2 x^4 + \frac{k}{2m} x^2 p_x^2 \qquad \text{(III-9)}$$

The difficulty arises from the term $x^2 p_x^2$. A partial symmetrization according to (III-5) gives

$$E^2 = \frac{1}{4m^2} p_x^4 + \frac{1}{4} k^2 x^4 + \frac{k}{4m}(x^2 p_x^2 + p_x^2 x^2) \qquad \text{(III-10)}$$

to which it corresponds the following operator

$$(\hat{H})^2 = \left(\frac{1}{2m} \hat{p}_x^2 + \frac{1}{2} k x^2\right)^2 \qquad \text{(III-11)}$$

Given the fact that any recipe which leads to an operator different from $(\hat{H})^2$, involves necessarily that $(\Delta E)^2$ is different from zero, the custom has prevailed to adopt the following rule

$$op(AB) = \frac{1}{2}(\hat{A}\hat{B} + \hat{B}\hat{A}) \qquad \text{(III-12)}$$

i.e. a symmetrization at the level of the two hermitian operators $\hat{A}$ and $\hat{B}$. Relation (III-2) is, of course, a particular case of (III-12).

Unfortunately, rule (III-12) is not able to be generalized. Let us consider a product ABC. In order to apply rule (III-12), we can, indeed, write either

$$(AB.C) = \frac{1}{2}(AB.C + C.AB) = \frac{1}{4}(ABC + BAC + CAB + CBA)$$

or $\quad (A.BC) = \frac{1}{2}(A.BC + BC.A) = \frac{1}{4}(ABC + ACB + BCA + CBA)$

In the general case, the operators which correspond to these expressions are different from one another, unless the operators $\hat{A}$, $\hat{B}$ and $\hat{C}$ commute, i.e. $\hbar$ is equal to zero. That is the Temple paradox [7] . This author, besides, admits that such a conclusion "destroys the whole structure of the modern form of quantum theory" !

Hence, rule (III-12) and, consequently, rule (III-2) which is a particular case of (III-12) , lead to a logical contradiction. Therefore, they have to be rejected. All the other recipes (III-4,5,6,...) lead to the same difficulty, except the complete symmetrization, e.g. (III-7) . Therefore, this procedure seems to be the sole one which has to be retained.

Energy fluctuation

The first consequence of the complete symmetrization of the operators is the existence of fluctuations of energy because $(\Delta E)^2$ is different from zero. We will give some examples.

For the harmonic oscillator, we obtain

$$op(E^2)_{sym} = (\hat{H})^2 + \frac{k}{4m}\hbar^2 \qquad\qquad \text{(III-13)}$$

Then

$$(\Delta E)^2 = \frac{k}{4m}\hbar^2 = (E_0)^2 \neq 0 \qquad\qquad \text{(III-14)}$$

E_0 being the energy of the ground state.

More generally [8] , by using the commutation relationship

$$[xp_x - p_x x] = i\hbar \qquad\qquad \text{(III-15)}$$

we see that

$$op(x^n p_x^2)_{sym} = \frac{1}{2}(x^n p_x^2 + p_x^2 x^n) + \frac{n(n-1)}{4m}\hbar^2 x^{n-2} \qquad \text{(III-16)}$$

so that, for a particle whose potential energy is of the following form

$$U(x) = \sum_n K_n x^n \qquad \text{(III-17)}$$

we obtain

$$op(E^2)_{sym} = (\hat{H})^2 + \sum_n K_n \frac{n(n-1)}{4m} \hbar^2 x^{n-2} \qquad \text{(III-18)}$$

From which it results

$$(\Delta E)^2 = \sum_n K_n \frac{n(n-1)}{4m} \hbar^2 \left\langle \psi x^{n-2} \psi \right\rangle = \frac{\hbar^2}{4m} \left\langle \psi \frac{d^2 U}{dx^2} \psi \right\rangle$$

$$= \frac{\hbar^2}{4m^2} \int \left(\psi''^2 + \frac{1}{3} \frac{\psi'^4}{\psi^2} \right) dx \qquad \text{(III-19)}$$

which is positive whatever U may be.

More generally, for a 3D-problem [8] , if the potential energy is of the following form

$$U(x,y,z) = \sum K_{n,r,s} x^n y^r z^s \qquad \text{(III-20)}$$

we obtain

$$(\Delta E)^2 = \frac{\hbar^2}{4m} \left\langle \psi . \nabla^2 U . \psi \right\rangle \qquad \text{(III-21)}$$

A particularly important case is that of the Coulomb potential for which, out of the charges, $\nabla^2 U = 0$. At the point where the charge Z is located, we have

$$\nabla^2 U = -4\pi\rho = 4\pi Z \delta(\mathbf{r}) \qquad \text{(III-22)}$$

For an atomic or molecular system, described by the Slater determinant

$$\psi = \left| \varphi_1 \overline{\varphi_1} ... \varphi_i \overline{\varphi_i} ... \varphi_n \overline{\varphi_n} \right| \qquad \text{(III-23)}$$

φ_i being an atomic or molecular orbital, the average value of the quadratic dispersion of the energy (in a.u.) is the following

$$(\Delta E)^2 = 2\pi \sum_K Z_K \sum_i \varphi_i^2(K) + 8\pi \left[\left\langle \left(\sum_i \varphi_i^2 \right)^2 \right\rangle + 2 \sum_{(ij)} (\varphi_i^2 \varphi_j^2) \right] \qquad \text{(III-24)}$$

(K = nucleus K , $\varphi_i^2(K)$ = values of φ_i^2 at the point where is located the nucleus K).

As an example, the following Table gives the values of ΔE and the average value $\overline{T}$ of the kinetic energy for various systems.

System	$\Delta E(a.u.)$	$\overline{T}(a.u.)$
Hydrogen ($1s$)	1	0.5
He^+ ($1s$) : $Z_1 = 2$	4	2
He $(1s)^2$: $Z_1 = 1.69$	4.9	2.9
$Be^{2+}(1s)^2$: $Z_1 = 3.69$	21.3	13.6
Be $(1s)^2(2s)^2$: $Z_1 = 3.69$; $Z_2 = 1.95$	22.8	14.6
H_2^+ ($R = 2.5 a.u.$; $1s$ - basis $Z_1 = 1.23$)	1.2	0.5
$(R \rightarrow \infty)^*$	1.0	
H_2 ($R = 1.4$ $a.u.$; $1s$ - basis $Z_1 = 1.19$)	2.0	1.1

* ΔE is maximum for the equilibrium distance.

Case of the operators associated with M_z^2 and M^2

Energy is not the unique property for which the complete symmetrization leads to a non-zero quadratic dispersion, so even when the system is in an eigenstate. Operators $\hat{M}_z^2$ and $\hat{M}^2$, associated with the square of the z - component of the kinetic momentum, and with the square of the momentum itself respectively, are affected by the complete symmetrization. Indeed, given that

$$\hat{M}_z = \frac{\hbar}{i}\left(x\frac{\partial}{\partial y} - y\frac{\partial}{\partial x} \right) \tag{III-25}$$

by symmetrizing completely with respect to the canonical variables q and

p [5,6,8] , we obtain

$$op(M_z^{\ 2})_{sym} = (\hat{M}_z)^2 + \frac{\hbar^2}{2} \qquad\qquad \text{(III-26)}$$

and, consequently, for the square of the kinetic momentum M^2

$$op(M^2)_{sym} = \left(\hat{M}_x\right)^2 + \left(\hat{M}_y\right)^2 + \left(\hat{M}_z\right)^2 + \frac{3}{2}\hbar^2 \qquad\qquad \text{(III-27)}$$

In other words, even if the wave-function is an eigenfunction of $\hat{M}_z$, the quadratic dispersion $(\Delta M_z)^2$ is different from zero. It is equal to $\hbar^2 / 2$. Likewise, for M^2, the result is different from that obtained by the orthodox quantum mechanics. In the hydrogen atom, for instance, M^2 is equal to $\left[\ell(\ell+1) + \frac{3}{2}\right]\hbar^2$ instead of $\ell(\ell+1)\hbar^2$. This significates that, even in the ground state $(\ell = 0)$, the value of M^2 is different from zero.

Another consequence of the complete symmetrization is that the energy of a rigid rotator is never equal to zero. Indeed, the Schrödinger equation which governs such a system is

$$\frac{1}{2I} op(M_z^{\ 2})\psi = E\psi \qquad\qquad \text{(III-28)}$$

The eigenvalues of which are

$$E_k = \frac{\hbar^2}{2I}(k^2 + \frac{1}{2}) \qquad\qquad (k = 0,1,...) \qquad\qquad \text{(III-29)}$$

Likewise, the rotation energy of a molecule whose principal moments of inertia are I_X, I_Y, I_Z , is shifted by $\dfrac{\hbar^2}{4}\left(\dfrac{1}{I_X} + \dfrac{1}{I_Y} + \dfrac{1}{I_Z}\right)$, and for a linear molecule by $\hbar^2 / 4I$.

Numerically, the effect is weak : 0.5 meV for $ClCH_3$,0.05 meV for benzen C_6H_6 , 4 meV for H_2, 0.5 meV for HCl, 0.01 meV for Cl_2, 0.05 meV for O_2 [9] .

Comparison with experiment

As we have seen, the complete symmetrization of the operators makes a certain number of divergences with respect to the conventional quantum mechanics. Although this rule is finally the most logical owing to the fact that it leads neither to an internal contradiction concerning the formalism, nor to any ambiguity, we have, nevertheless,to wonder whether this rule is consistent with experiment.

Concerning the kinetic momentum operators on which the rotation spectroscopy is based, we have seen that, contrary to the assertion of the conventional quantum mechanics, the rotation energy is never equal to zero. Nevertheless, for a given system, e.g. in the plane rotator (III-26) as well as in a molecule (III-27), all the rotation levels are shifted by the same value, so that the spectrum is unchanged. Likewise for the various thermodynamical properties deduced from the partition function, the specific heats, for instance. On the contrary, the existence of a residual rotation energy ($\hbar^2 / 4I$) could have an influence on the dissociation energy. At the present time, the accuracy obtained for the molecules other than H_2 (e.g. 8 meV for O_2) precludes any conclusion. For H_2, the test seems to be negative, the residual energy is, indeed, equal to 4 meV whereas the announced experimental accuracy is equal to 0.03 meV [9] . In fact, the problem is more complex. The dissociation energy, indeed, is obtained by extrapolation of the various vibration levels by taking the anharmonicity and the rotation-vibrational interactions into account. It would be necessary to take up the problem again by introducing a correction, specific of each of the levels respectively.

The continuous rotation of a molecule could also to play a role in another domain, e.g. in that of the dielectric properties. The average value of the component of the moment of a polar molecule, which is equal to zero along a given direction, is interpreted as arising from the shocks which disorient the molecules. In other words, another reason would be the cause of the phenomenon. Unfortunately, the theory of the interactions within the liquids is too rudimentary to hope for infering some conclusion concerning the difference between the two models.

In any case, we must quite admit that it is not absurd to find a never equal to zero energy for the rotator from the moment when all the theorists are unanimous in saying that the oscillator is never at rest. The physical cause which induces the unceasing motion of the oscillator must forbid the rotator to be at rest. In this connection, we will remark that the average quadratic dispersion of M_z^2 in the plane rotator $\left(\psi = \dfrac{1}{\sqrt{2\pi}} e^{im\varphi}\right)$ is

$$(\Delta M_z)^2 = \overline{M_z^2} - (\overline{M_z})^2 = \left\langle \hat{M}_z^2 + \frac{\hbar^2}{2} \right\rangle - \left\langle \hat{M}_z^2 \right\rangle = \frac{\hbar^2}{2} \qquad \text{(III-30)}$$

Given that, on the other hand, for the angular coordinate φ, we have

$$(\Delta\varphi)^2 = \frac{1}{2\pi} \int_{-\pi}^{\pi} \varphi^2 d\varphi = \frac{\pi^2}{3} \qquad \text{(III-31)}$$

we obtain

$$(\Delta M_z)(\Delta\varphi) = \frac{\pi\hbar}{\sqrt{6}} = 0.72\hbar \qquad \text{(III-32)}$$

Now, in the conventional quantum mechanics, the general relationship

$$(\Delta M_z)(\Delta\varphi) \geq \frac{\hbar}{2} \qquad \text{(III-33)}$$

is not verified owing to the fact that $\Delta M_z = 0$. This remark is a supplementary argument for the complete symmetrization of the operators.

Let us quote a last example of result more logical than that admitted by the conventional quantum mechanics, namely that of the square of the kinetic momentum in the hydrogen atom. In its ground state, according to the orthodox formalism, the value of this property is equal to zero, which would signify that, either the electron is at rest, or it is moving along a straight line passing through the nucleus and keeping a fixed direction in space. These two conclusions are physically absurd : Indeed, for the first one, the average kinetic energy is different from zero, and for the second one, $\overline{x^2} = \overline{y^2} \neq 0$. Dirac [10] who was conscious of the difficulty, in 1930, thought that $\hat{M}_x{}^2 + \hat{M}_y{}^2 + \hat{M}_z{}^2 = \hat{\beta}$ was not the correct operator associated to M^2, and that it was better to use $\hat{\beta} + \dfrac{\hbar^2}{4}$. In our model, we must adopt $\hat{\beta} + \dfrac{3}{2}\hbar^2$. Whatever that may be, the Dirac idea has been quickly forgotten, probably because it was inconsistent with the rule (III-2)

Concerning $(\Delta E)^2 \neq 0$ and the energy fluctuations which this result involves, the situation, at first sight, appears as being much less favorable, given that the absorption spectrum is formed by sharp lines. In fact, we are prisoners of the interpretation which has been proposed at the beginning of this century, and which strongly connects the discrete absorption spectrum with the existence of well-defined energy levels . The absorption spectrum is the response to an external electromagnetic excitation and we have no reason to claim that a continuous distribution in energy cannot lead to discrete spectra. As a support of our assertion, we will quote the case of the harmonic oscillator, for which stochastic electrodynamics, in spite of the continuous energy distribution, obtains a discrete spectrum. In any case, the spectrum lines are not infinitely sharp, as one is too oft inclined to say, so that the objection is not very serious.

On the contrary, the energy fluctuations allow to explain the penetration of the particle in classically forbidden domains and its passage through potential barriers whose height is superior to its energy (tunnelling effect) , phenomenons which are considered as being typically quantal. Let us consider an harmonic oscillator whose potential energy is equal to $U = \dfrac{1}{2}kx^2$. The expression of the total energy $E = \dfrac{1}{2}mv^2 + \dfrac{1}{2}kx^2$, shows that if E is constant (equal to E_{AB}), the particle is moving on the segment $A'B'$ $(\pm\sqrt{2E_{AB}/k})$. If E fluctuates from zero to infinity, all the values of x can be reached, so that, if the observed value is conventionally considered as being constant, equal to E_{AB} , we are then constrained to assign a surprising power to the particle, namely that of penetrating inside forbidden domains ($x < x_{A'}$, $x > x_{B'}$) for which the kinetic energy is negative. Likewise for the tunnelling effect which can be explained by the fact that the particle does not pass through the barrier but it is jumping over this latter more or less easily according to the energy which it receives from the universe field. Consequently, in these two cases, it is not necessary to put forward any quantal behavior, the classical concepts being quite valid.

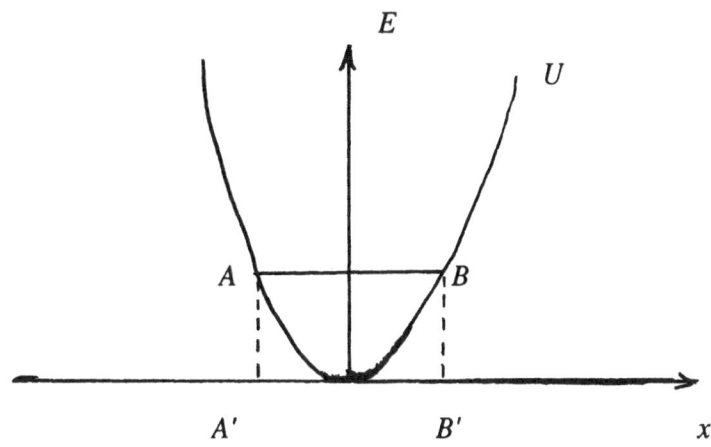

Moreover, the energy fluctuation explains why a particle is partially reflected by a potential barrier whose height is weaker than the energy of the incident particle. Through the fluctuations indeed the energy of the latter falls at times below the top of the barrier, which forbids the particle to proceed on one's way. The greater is the ratio energy of the particle / height of the barrier, the weaker is the effect, which agrees with the conventional quantum result [11] .

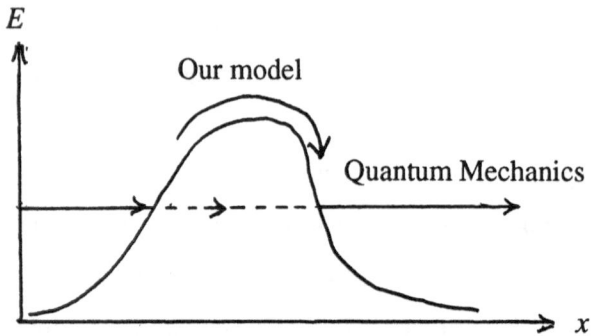

Behavior of a particle coming in front of a potential barrier : According to Quantum Mechanics, the particle passes through the barrier by tunnelling effect, its energy remaining constant. In our model , the particle jumps over the barrier when it receives a sufficiently great energy from the universe field.

Conclusion

It results from this discussion that, far from being cumbersome, the complete symmetrization of the operators, not only leads to any inconsistency with experiment, but also allows a more logical physical understanding of the observed phenomenons. In any case, the most important fact which has to be retained, is the fluctuation in energy which our model just forecasts (Chap. II) .

A question occurs naturally to us. Do systems exist in which such a fluctuation can be detected ? The spontaneous racemization we are going to examine in the following chapter, will bring useful information about this problem.

References

[1] E. Schrödinger, *Ann. Physik* **79** (1927) 734.

[2] H. Weyl, *The theory of groups and quantum mechanics* (Dutton, New York) 1931, J.E. Moyal, *Proc. Cambridge Phil. Soc.* **45** (1949) 99.

[3] H. Margenau, R.N. Hill, *Proc. Theoret. Phys. (Kyoto)* **26** (1961) 722.

[4] M. Born, P. Jordan, *Z. Phys.* **34** (1925) 873.

[5] T.H. Boyer, *Phys. Rev.* **D 11** (1975) 809.

[6] A. Julg, *Theoret. Chim. Acta* **74** (1988) 323; in *Courants, amers, écueils en microphysique,* C. Cormier-Delanoue, G. Lochak and P. Lochak edn. (Fondation Louis de Broglie, Paris) 1993, p. 211.

[7] G. Temple, *Nature* **135** (1935) 957.

[8] A. Julg, P. Julg, *Folia Theoret. Chim. Latina* **7** (1979) 63.

[9] P. Huber, G. Herzberg, *Molecular spectra and molecular structure . IV. Constants of diatomic molecules* (van Nostrand-Reinhold, Princeton) 1979.

[10] P.A.M. Dirac, *The Principles of Quantum Mechanics* (Clarendon Press, Oxford) 3rd edn., 1947, p. 144.

[11] I.I. Goldman, V.D. Krivchenkov, V.I. Kogan, V.M. Galitski, *Problems in Quantum Mechanics,* translated, edited and arranged by D.Ter Haar (Infosearch Ltd, London) 1960. French translation by R. David, *Problèmess en Mécanique Quantique* (Masson, Paris) 1969, p. 130.

IV

The problem of enantiomers

The classical point of view

Classically, a molecule is considered as built up upon nuclei linked by electron pairs (Lewis [1]) . A simple mechanical model consists of comparing the electron pairs with coil-springs which allow longitudinal and angular vibrations around the equilibrium positions of the nuclei. This model, in spite of its simplicity, permits, on the one hand, explanation of molecular conformations as revealed by experiment (X-rays, electron diffraction ,...) , and, on the other, interpretation of vibration spectra (infra-red and Raman) in a very satisfying manner.

The oscillations of nuclei around their equilibrium positions being generally of small amplitude, the topology of the molecule is not affected by the vibrations, so that the latter can be neglected in defining the chemical species under consideration. Moreover, it is well known that, following the works of Pasteur and those of Le Bel and Van't Hoff on crystal and molecules, if a molecular structure - defined as a set of points at rest (the nuclei) - cannot be superposed on its image, it exhibits a rotatory power. Two structures symmetrical with respect to a mirror and not superposable on their image possess opposite rotatory powers. They are called *enantiomers* . Two enantiomers (A^+ and A^-) exhibit the same scalar properties (density, melt-point, frequencies and modes of vibration, etc.) . Nevertheless, these structures have to be considered as two different molecules, each playing the role of an independent component in the phase rule (existence of an eutectic A^+A^-) . Two enantiomers react similarly on inactive molecules to give compounds which are also enantiomers. On the contrary, with an optically active substance (e.g. : B^+), A^+ and A^- give two different compounds, A^+B^+ and A^-B^+, called *diastereoisomers* , which have different scalar properties, allowing us to separate them.

L-form   D-form

An example of enantiomeric molecules : The L- and D-forms of amino acids (R is a more or less complex chain , e.g. CH_3 in alanine). The symbols L and D respectively refer to the *levo-* and *dextro-* rotatory forms of glyceraldehyde $HOCH_2CH(OH)CHO$ from which formally the form under consideration arises. The fact that the molecule belongs to the series L (or D) does not involve that it is levo- (or dextro) rotatory. For the greatest number of amino acids, the L-forms are dextrorotatory and, conversely, the L-forms levorotatory. The proteins of the living organisms are built up upon L amino acids exclusively.

The case of asymmetric amines $NR_1R_2R_3$ sets a problem. Although these molecules can be conceived in two enantiomeric forms, it is impossible to separate these forms. No optical activity is observable. As far back as the beginning of the twentieth century this impossibility has been interpreted as arising from a very fast oscillation of the nitrogen nucleus with respect to the average plane of the ligands, the life-time of each isomer being too brief to permit reaction on another active molecule or the observation of any optical activity.

The similar case of arsines $AsR_1R_2R_3$, discovered subsequently, supports this interpretation. These molecules, indeed, can be separated into two enantiomers [2] . Nevertheless, a given enantiomer cannot be indefinitely stored in its pure form. After a certain number of months it becomes racemized, i.e. transformed into an equimolecular mixture of both enantiomers. Phosphines $PR_1R_2R_3$ present the same phenomenon, but the racemization is faster and only a few weeks are necessary [3] .

Classically, the passage from one form to the other can be conceived only by an intermediate structure in which the three bonds carried by the apical atom (N, As or P) are coplanar. The ease of the transformation of one form into the other depends on the height of the potential barrier which divides the enantiomers in their equilibrium position. The higher the barrier, the greater the life-time of each enantiomer. In the case of molecules whose asymmetry center does not carry three ligands as in the pyramidal molecules we have considered, but four ligands, (e.g. : amino-acids $H_2N - CH(R) - CO_2H$, R being a more or less complex chain), the transformation is very difficult. It requires, indeed, an energy of about 100 kcal per mole, whereas 6, and 30 to 40 are sufficient for amines and arsines respectively. So the racemization is not observable at the human scale, each enantiomer appears as being infinitely stable. The racemization, nevertheless, can be observed over a larger time, for instance, at the geological scale, for the amino-acids of collagen in fossil bones, so that the racemization rate of these amino-acids is used for dating the fossil bones [4] .

A point, nevertheless, remains mysterious concerning such a classical interpretation. Even in amines where the lowest barrier occurs, the corresponding energy is greater than the thermal energy at the ambient temperature, so that the jump from one form to the other appears incomprehensible within the classical interpretation.

The quantal point of view

The problem of enantiomers in quantum mechanics, raised by Hund [5] as far back as 1927, is, in fact, a particular case of the more general problem of the molecular structure we will study in Chap. VI. At present, we will be content with the traditional approach which formally reduces the problem of enantiomers to that of a particle in a symmetric double-well potential. We will neglect the weak neutral current perturbations which, owing to the parity non-conservation, provoke a very small asymmetry for the double well : $\Delta E \cong 10^{-20} a.u.$ for an amino-acid [6] (More recent calculations [7] give a greater difference in energy : $12.3 \ 10^{-20} a.u.$ between the L and D forms of alanine).

Consequently, we have to solve the time-independent Schrödinger equation

$$\hat{H}\psi = E\psi \qquad \text{(IV-1)}$$

where $\hat{H} = \hat{T} + V$, with $V(-x) = V(x)$.

In the general case, it is impossible to integrate this equation. Nevertheless, for a qualitative discussion the exact form of V is not very important, so that we can use approximate potentials.

Practically, one often uses piecewise-continuous potentials, or continuous with discontinuous first derivatives. Let us quote for instance

$$\begin{cases} V = \infty & \text{if } x < -a \text{ and } x > a \\ V = V_0 & -b < x < b \\ V = 0 & -a < x < -b \text{ and } b < x < a \end{cases} \qquad \text{(IV-2)}$$

or

$$\begin{cases} V = \dfrac{1}{2}k(x+a)^2 & (x < 0) \\ V = \dfrac{1}{2}k(x-a)^2 & (x > 0) \end{cases} \qquad \text{(IV-3)}$$

In the first case, the solution is obtained by joining the solutions corresponding to the various domains. In the second case, one builds up the solution as linear combinations of eigenfunctions of the two harmonic oscillators respectively centred at $+a$ and $-a$. These functions are twofold degenerate. The diagonalization of the energy-matrix corresponding to the actual potential splits up the degenerate levels. One obtains pairs of simple levels, alternatively symmetrical and antisymmetrical. Practically, in order to obtain the first two levels, one contents oneself with the following combinations

$$\begin{cases} \psi_S \propto (\phi_1 + \phi_2) \\ \psi_A \propto (\phi_1 - \phi_2) \end{cases} \tag{IV-4}$$

ϕ_1 and ϕ_2 corresponding respectively to the ground state of each oscillators (IV-3).

Among the continuous potentials the best known is the one proposed by Manning [8]

$$V = -A\,\mathrm{Ch}^{-2}\left(\frac{x}{a}\right) + B\,\mathrm{Ch}^{-4}\left(\frac{x}{a}\right) \tag{IV-5}$$

The solution can be expressed by means of the roots of a continued fraction.

More recently [9] we have proposed the following potential

$$V = \frac{k}{2}\left[\frac{(x+a)^2}{1+e^{4\alpha ax}} + \frac{(x-a)^2}{1+e^{-4\alpha ax}} \right] \tag{IV-6}$$

to which corresponds, for the ground state, the following function

$$\psi \propto e^{-\alpha(x-a)^2} + e^{-\alpha(x+a)^2} \tag{IV-7}$$

i.e. the sum of two Gaussian functions.

In all the cases, whatever the used potential may be, the qualitative results are the same, namely :

i. the wave-function corresponding to the ground state is x-symmetrical, consequently the rotatory power is equal to zero,

i.i. one obtains a spectrum of frequencies connected to the motion of the nuclei,

i.i.i. the difference in energy between the first levels, $E_A - E_S$, is directly connected with the height of the potential barrier.

Consequently, on the one hand, the quantal result concerning the optical activity disagrees with experiment because the arsines and the phosphines are optically active, and, on the other, no information is obtained concerning the actual dynamics of the system.

In order to attempt to find again the classical concept of oscillation of the apical nucleus, one argues as follows : One starts from a distribution strongly concentrated above a well, ϕ_1^2 for instance. Then, using the time-dependent Schrödinger equation, one follows the evolution of the density versus time. This density exhibits a maximum which oscillates between the two wells. Practically, by introducing only the first states ψ_S and ψ_A (IV-4), the wave-function is as follows

$$\Psi(x,t) = C_S \psi_S(x) e^{-iE_S t/\hbar} + C_A \psi_A(x) e^{-iE_A t/\hbar} \qquad \text{(IV-8)}$$

The coefficients C_S and C_A being practically equal ($E_A - E_S$ is very small), the density is equal to

$$|\Psi^2| \approx \psi_S^2 + \psi_A^2 + \psi_S \psi_A \cos\left[(E_A - E_S)t/\hbar\right] \qquad \text{(IV-9)}$$

The oscillation period is precisely equal to $h/(E_A - E_S)$, i.e. corresponds to the $\psi_S \to \psi_A$ transition.

This result is generally considered as justifying the classical concept of oscillation of the apical nucleus between the two wells. In connection with this interpretation, it is funny to remark that quantum mechanics considers the planar rotator as being at rest in its ground state, though a wave-packet initially centred at a given point is periodically deformed. The wave-packet spreads, afterwards it concentrates above another point, and so on.

In any case, this way of presenting the situation is very questionable. The actual problem, indeed, is that of the ground state of the system. Now this state is a stationary state. Its density is time-independent. Consequently it presents no oscillation from one well to the other, even if the nuclei oscillate effectively. Function (IV-8) does not corresponds to a stationary state. The study of the evolution of a wave-packet initially concentrated above a well, is a problem completely different from our problem. Given that the wave-packet does not correspond to a stationary state, quantum mechanics defines the average energy only for such a state. This energy is greater than the energy of the ground state. Thus, according to the orthodox interpretation, the oscillation is not an observable phenomenon.

Independently of the fact that the rotatory power of the system in its ground state is equal to zero, another ticklish point has to be emphasized. The integration of the Schrödinger equation (IV-1) leads to a wave-function which describes both the two enantiomers . Consequently, a same function would describe two different molecules ! In any case, to say that each of the two density maxima $|\psi|^2$ corresponds to one of the two enantiomers respectively, is an error . Schrödinger equation does not describe a chemical mixture.

Another possible approach consists in considering that the system is not isolated but submitted to a perturbation potential arising from the surrounding medium. Then, the corresponding Schrödinger equation is non-linear

$$\hat{H}\psi + V[\psi] = E\psi \qquad\qquad\qquad (IV\text{-}10)$$

From the latter, it is easy to show that, starting from the "mixture" $\phi_1 + \phi_2$, under the effect of a very small perturbation, the system can rock into the one or the other well. In other words, we have a *breakdown* of the symmetry which leads to a localized state [10] .

The Monte-Carlo method, applied to a flow of two states systems, which, here, would correspond to ψ_S and ψ_A, leads to an analogous result. Under the effect of the interaction between the systems, the set of the systems rocks into a localized state.

Although these interpretations, based on the orthodox quantum mechanics, explain the possibility to obtain of the two chemical forms, it remains to understand why the system does not remain indefinitely in the well in which it is fallen.

Our interpretation

As we have seen in the previous chapter, the passing of a particle through a potential barrier whose height is greater than the average energy of the particle does not raise difficulties within our model, because, under the random variations of the universe field, the particle is able to acquire an energy sufficient for *jumping* over the barrier.

Let us consider a certain number N of molecules of the same configuration (e.g.:d) , these molecules which, according to their history, exhibit different energies distributed around their average value, either remain within the same potential well, or randomly pass into the other well, so that, after a sufficiently long time, we have $N/2$ molecules of form d and $N/2$ of form l . In other words, the substance is racemized. We will notice that, contrary to the approaches based on the bifurcation theory, the system does not remain confined within one of the wells. The apparent difference in the behavior of the various chemical compounds able to present such a phenomenon, arises from the racemization rate only, i.e., lastly, from the ease of passing from one form to the other. This rate which is too great, at our scale, for amines (the inversion requires $10^{-12}s$ only) , does not allow the separation of the two forms and the observation of the jumps. On the contrary, for arsines, the two inverse forms are observable and behave as distinct molecules. For each of the latter, we can define an equilibrium geometry, an electron density, an absorption spectrum, a rotatory power. Nevertheless, the question to know the true meaning which has to be ascribed to the word "molecule", and that of the density deduced from the Schrödinger equation for the d and the l isomers considered as a whole, are not settled for all that.

In fact, this problem is a particular case of the more general question of the connection which exists between the quantal forecasts and the reality, or, at least, the knowledge we can reach about this connection. That will lead us to hark back to the measurement itself.

Analogy with ferromagnetism

Although the phenomenon is more complex owing to the fact that it implies a great number of electrons, the case of ferromagnetism presents an analogy with the one of the existence of enantiomeric molecules of non zero rotatory power which are stable during a more or less long time. Indeed, ferromagnetism arises from the parallelism between the spins of free electrons of the magnetized body. With respect to the latter, these spins can be oriented along two opposite directions, and thereby create opposite magnetic fields. Now according to Quantum Mechanics, these two configurations have to be taken in the wave-function into account with equal weights so that the resulting field should be equal to zero. Thus the situation is quite comparable with the one we have encountered for arsines. In the present case, the number of spins which have to be inversed to pass from a configuration to another is extremely great owing to the size of the magnet, which entails that the ergodicity time can be considered as being practically infinite. From which the existence of permanent magnets.

References

[1] G.N. Lewis, *J. Amer. Chem. Soc.* **38** (1916) 762.

[2] R.E. Weston, *J. Amer. Chem. Soc.* **76** (1954) 2645.

[3] C.C. Costain , G.B.M. Sutherland, *J. Phys. Chem.* **56** (1952) 321.

[4] J.L. Bada, R. Protsch, *Proceed. Nat. Acad. Sc. USA* **70** (1973) 1331 ;

A. Julg, R. Lafont, G. Périnet, *Quaternary Science Rev.* **6** (1987) 25.

[5] F. Hund, *Z. Physik* **43** (1927) 805.

[6] S.F. Mason, G.E. Tranter, *Chem. Phys. Lett.* **94** (1983) 34 ;

G.E. Tranter, *Mol. Phys.* **56** (1985) 825.

[7] P. Lazzaretti, R. Zanasi, *Chem. Phys. Lett.* **279** (1997) 349.

[8] M.F. Manning, *J. Chem. Phys.* **3** (1935) 136.

[9] A. Julg, *Croatica Chem. Acta* **57** (1984) 1497.

[10] P. Claverie, G. Jona-Lasinio, *Phys. Rev.* **A 33** (1986) 2245 ; *Progress of Theoret. Physics* , Suppl. **86** (1986) 54.

V

Measurement of a property and ergodicity time

A preliminary remark

If there are no doubt that the knowledge we can have about a system arises from the perception we can obtain concerning it by performing an experiment, either direct if we appeal to our senses, or indirect if we use a measurement device, we cannot claim, just as the Copenhagen School does, that any physical object does not have an own existence outside the knowledge which the observer can have concerning it. The situation would be the same as if we claim that America did not exist before Europeans discovered it !

In fact, the problem is the following. When we measure the property G of a system described by the wave-function $\psi = \sum C_i \phi_i$, ϕ_i being an eigenfunction of $\hat{G}$, the result of the measurement is one of the eigenvalues of $\hat{G}$, e.g. g_k , with a probability equal to $\left| C_k \right|^2$. For the Copenhagen School, the effect of the measurement would be to reduce the function $\psi = \sum C_i \phi_i$ to one of its components ϕ_k, i.e. to provoke the pasage of the system from the initial state described by ψ to a certain state ϕ_k in which the property G has a well-defined value. Consequently, before the measurement being performed, it is stated that we have no precise knowledge concerning the system, which precludes us to say anything about the state of the system, and, by going to extremes, about the existence of the system.

Nowadays, this dogmatic point of view turns out obsolete because of the advent of more comprehensible approaches involving the development of the theories of decoherence by Zurek and consistent histories by Griffiths [1] . Anyhow, the role played by the measurement in the Copenhagen interpretation is not arguable for many reasons. Neither the measurement device nor the observer appear nowhere in the equations of quantum formalism [2] . On the other hand, whatever the device, the experimentalist, the place and the date of the experiment may be, the same properties are obtained for a given system. Relations, as that of Heisenberg, $\Delta x . \Delta p_x \geq \hbar / 2$ for instance, refer to the system itself, and not to the device or the procedure of the measurement. All the dissertations on the role of the measurement device are, in fact, according to the expression of Bunge [2] , a *philosophical appendix* only which does not bring nothing to the quantum formalism.

The story of the Schrödinger cat quite illustrates the standard manner of interpreting the quantum formalism. The poor cat, kept in an opaque box with a murderous apparatus which has one chance over two of having killed it after a certain time, is, for the Copenhagen School, neither dead or living, as long as we have not opened the box, i.e. performed the experiment. Given that the probability to find the cat living (or dead) is equal to 1/2, we can consider that the system, i.e. the cat, is described by the following wavefunction $\psi = 2^{-1/2} (\phi_{\text{liv.}} + \phi_{\text{dead}})$, $\phi_{\text{liv.}}$ and ϕ_{dead} being the two eigenfunctions of the operator " life or dead". The cat is thus both half-dead and half-living. In other words, we do not know the actual state of the cat. When we open the box, we find the cat either dead or living, i.e. in a state described either by ϕ_{dead} or by $\phi_{\text{liv.}}$.

Within a statistical framework, the interpretation of this experiment is completely different. Indeed, if we do not consider one cat only, but a very great number of cats within an equal number of murderous boxes, before performing the experiment, i.e. before opening all the boxes, we can assert that one half of the cats is dead, the other living. The measurement for one cat effectively gives a dead cat or a living one with equal probabilities. But, in no case, the fact that we have opened the box, is the cause that the half-dead cat has been completely killed if we find the cat dead, and conversely that the half-living cat has been half-revived if we find it living.

Very recently, a device mimicking the conditions of the cat Gedanken-Experiment at the atomic level has been achieved in Paris [3] : It involves a Rydberg atom of rubidium as a two-state microscopic system and a resonating cavity of photons as a measurement apparatus; this allows to see the quantum superposition of states and its annihilation after a very short time because of the effect of destructive interferences due to the coupling with environment, that is to say the so-called *decoherence* [4] . In other words, under the effect of an external perturbation, very quickly the cat in the box would become living *or* dead instead of being living *and* dead.

Whatever that may be, for our part, we think that the system does effectively exist, independently of any observation, and that the quantum formalism is only a practical tool to know the result of a measurement of a given property of the system, independently, of course, of any physical perturbation, often difficult to avoid.

Physical meaning of the measurement result

Any physical process, so quick it is, cannot be considered as being instantaneous. It always demands a certain time. The measurement of a property is submitted to this necessity. The indication given by the measurement device, is only the average value of the measured property over the measurement time τ_m. For instance, the pressure indicated by a manometer immersed within a gas corresponds to the average value of the forces due to the shocks of the molecules against the manometric membrane. The pressure seems to us constant because the impacts of the molecules are too close together with respect to the response-time of the apparatus. If we could use a membrane which would respond more quickly with regard to the impacts of the molecules, the result would be different. We would obtain values which would fluctuate around the previous average value.

Another typical example is that of the Newton disk. Let us divide a disk into sectors colored according to the seven rainbow colors. When this disk is rotating about its axis with a sufficiently great angular velocity, it seems to be white. Why ? Because our eye integrates the perceptions received over a time interval of about 0.1 second. On the contrary, if we take photographs with an exposure of 0.001 second, we will see the various sectors of the disk at rest and colored with their respective color. Here still, the result of the measurement depends on the observation time.

More generally, if the studied system exhibits an ergodic behavior, i.e. that the average value of its various properties becomes stable after a sufficiently long time τ_e (the ergodicity time) , the result depends on the respective values of τ_e and τ_m. If $\tau_m > \tau_e$, we will directly obtain the limiting average value, and, consequently successive measurements will give the same result. On the contrary, if $\tau_m < \tau_e$, successive measurements will give values different from one another, and, in order to obtain a stable average value, a great number of measurements will be necessary. In any case, a fundamental point has to be emphasized, namely that the apparent difference in the behavior we could observe, does not correspond to any physical reality, but it arises from our observation techniques.

Some arguments for an ergodic interpretation of quantum mechanics

The previous remark can afford an argument in favor of the ergodic character of quantum mechanics. We have indeed seen in Chapter IV that, although the problem of amines and that of arsines are theoretically the same for both these families of molecules, namely that of a particle within a symmetrical double-well potential, these molecules exhibit very different behaviors: According to the quantal forecast, the former are not dedoublable, and their optical activity is equal to zero, whereas the latter are dedoublable into active antipodes, racemizing very slowly.

Let us assume that it is possible to observe a well-determined molecule of arsine and to draw the histogram of the values successively obtained for a given property. Owing to the oscillation between the two wells of potential corresponding to the two enantiomers respectively, over a sufficiently long time interval we will obtain a zero average value for the rotatory power and an average density of presence for the nucleus As, symmetrical with regard to the average plane of the ligands. On the contrary, for an amine, the oscillation is so rapid with respect to our present techniques, that we obtain the average value directly, in particular a rotatory power equal to zero. If for the density $|\psi|^2$ corresponding to the apical nucleus we do not obtain a distribution with two maxima corresponding to the two enantiomers respectively, the reason is that experiment does not distinguish the two forms from one another.

We can also quote the example of the electron densities obtained for atoms by X- or electron-diffraction, which coincide completely with the quantum forecast. These densities indeed arise from the integrating over the position of the electrons in an experiment which is sufficiently long for the average limit value to be reached.

On the contrary, the absorption spectra whose obtaining is practically instantaneous, can lead to different results. For instance, for some complexes of transition elements, the structure deduced from optical spectra is different from that deduced by X-spectrography [5] . The case of the perfluorocyclobutan C_4F_8 is analogous. This molecule appears as being planar in electron-diffraction and bent in infrared [6] . Techniques , as the Mössbauer effect and NMR, which utilize intermediary times, give results in accordance with the latter.

At last, let us recall the case of the diamagnetic susceptibility of an atom - say an hydrogen atom - which is proportional to $\overline{r^2} = 3a^2 / Z^2$, and which, in spite of a great quadratic dispersion $\Delta(r^2) = \sqrt{3/2}\,\overline{r^2}$, can be measured within a great accuracy, which clearly shows the rapidity of the fluctuation with respect to the measurement time.

To conclude, quantum mechanics in fact gives *the average value of the various properties over a sufficiently long time* [7] . Its character would thus be essentially *ergodic* . Moreover, we understand why in the Schrödinger equation commonly used for isolated systems, the *time* does not appear, in spite of the obvious dynamical character of this equation.

Estimation of the ergodicity time : A quantum approach

The interpretation we have above proposed is based on the value of the ergodicity time of the system under consideration. Before continuing the discussion, it is therefore necessary to obtain if not its exact value, at least an order of magnitude of this ergodicity time, knowing that for the quasi-totality of atomic and molecular systems whose behavior is well described by quantum mechanics, this time must be extremely small compared to our present investigation techniques, i.e. roughly $10^{-12}\,s$, the case of arsines and phosphines being an exception. But it is not excluded that in a near future, new systems will become accessible to our measurements, which would permit to test our interpretation as it is now the case for the Schrödinger cat paradox [3]

In order to estimate the ergodicity time, we can utilize the well-known relation of Heisenberg concerning the energy

$$\delta E.t \cong \hbar \qquad (V-1)$$

Although this relationship is one of the most famous results of quantum mechanics, its physical meaning is far from being clear. Quantum mechanics, indeed, considers the energy of an isolated system as being constant. But for Heisenberg, during the measurement, the system interacts with the measurement device so that, in fact, it cannot be considered as being isolated. Its energy must fluctuate. A measurement performed at the time t (computed from the beginning of the interaction between the system and the measurement device) gives a value different from that corresponding to the system in an isolated state, E_0, namely

$$E = E_0 + \delta E \qquad (V-2)$$

One proves [8] that the most probable value of δE is precisely given by relationship (V-1) .

Let us apply this relationship to our model by considering that the perturbation does not arise from the device, as the orthodox interpretation claims, but from the systems which surround the one we are studying. This relationship gives the instantaneous value of the energy at the time t . Let us perform successive measurements over the finite time interval $(0,\tau)$. According to the Schwarz inequality, we can write

$$\left| \int_0^\tau (E - E_0)t \, dt \right|^2 \le \int_0^\tau (E - E_0)^2 \, dt . \int_0^\tau t^2 \, dt$$

i.e., according to (V-1)

$$\hbar^2 \tau^2 \le \tau.(\Delta E_\tau)^2 . \frac{2}{3}\tau^3$$

$(\Delta E_\tau)^2$ being the quadratic dispersion observed for the energy over the time interval $(0,\tau)$. From which, after simplification,

$$\Delta E_\tau . \tau \geq \hbar \tag{V-3}$$

We will notice that this relationship is valid whatever the time origin may be, owing to the ergodic character of the system.

If we admit that quantum mechanics gives the average value over a sufficiently long time and that this value becomes stable if this interval is greater than the ergodicity time, we can consider that the latter is approximately given by

$$\tau_e \cong \hbar / \Delta E \tag{V-4}$$

i.e. $\quad \tau_e(s) \cong 2.10^{-17} (\Delta E_{a.u.})^{-1} \tag{V-5}$

ΔE being the value obtained from the complete symmetrization of the operators (Chap. III) .

An interesting point is worthy of notice : Contrary to the direct application of the definition of ergodicity we have given in Chapter II, this approach based on the Heisenberg relationship, does not refer to the accuracy demanded for the stabilization of the value of the property under consideration.

Orders of magnitude of the ergodicity time

The simplest case is that of the harmonic oscillator. We have, indeed, seen that $\Delta E = E_0$ (III-12) . Consequently, according to (V-4)

$$\tau_e \cong \frac{T}{\pi} \tag{V-6}$$

T being the classical period of the oscillator. For molecular systems, this time is thus of about $10^{-12} s$.

For a rigid planar rotator where $\Delta E \cong \hbar^2 / I$ (III-29) , we obtain

$$\tau_e \cong I / \hbar \tag{V-7}$$

i.e. $10^{-9} s$ for molecular rotations.

Another interesting case is that of a particle in a symmetrical double-well potential. In this formal discussion, we will use the model potential (IV-6) for which the exact solution for the ground state is the following

$$\psi(x) = \left(\frac{\alpha}{2\pi}\right)^{1/4}\left[e^{-\alpha(x-a)^2} + e^{-\alpha(x+a)^2}\right](1 + e^{-2\alpha a^2})^{-1/2} \qquad \text{(V-8)}$$

with $\alpha = (V_0/2a^2)E_0$, $E_0 = (\hbar/2)\sqrt{k/m}$ and $V_0 = V(0) = \frac{1}{2}ka^2$, $2a$

being the distance between the two minima.

If we assume that the ratio V_0/E_0 is great compared to unit, the height of the potential barrier is of the order of V_0. A straightforward calculation leads to

$$\Delta E \approx E_0\sqrt{\alpha a^2}e^{-\alpha a^2} = E_0\sqrt{V_0/2E_0}e^{-V_0/2E_0}$$

i.e. $\quad \tau_e \propto (V_0/2E_0)^{-1/2}e^{V_0/2E_0}$ \qquad\qquad\qquad (V-9)

If α tends to infinity, the potential (IV-6) can be compared to $kx^2/2 + V_0\delta(x)$. The energy of the corresponding ground state is equal to $3E_0$. In the general case, the value of the energy W_0 will be thus located between E_0 and $3E_0$. In order to obtain the order of magnitude of W_0, we will adopt the intermediary value $2E_0$. Consequently, according to formula (V-9), we obtain

$$\tau_e \propto (V_0/W_0)^{-1/2}e^{V_0/W_0} \qquad\qquad\qquad \text{(V-10)}$$

This relationship shows that the ergodicity time τ_e depends essentially on the ratio V_0/W_0, increasing very quickly as the ratio V_0/W_0 increases.

In NH_3 and amines, $V_0 \cong 5\ kcal/mol$ and $W_0 \cong 1\ kcal/mol$, i.e. $V_0/W_0 \cong 5$. In arsines, this ratio is of about 50. Consequently, the ergodicity time is multiplied by a factor of about 10^{19}. Now, for the harmonic oscillator we have seen that the ergodicity time is proportional to the period. If for the ergodicity time in an amine we adopt the value usually admitted for the oscillation period between the two wells, i.e. $10^{-12}\ s$, for an arsine, we obtain a time of about $10^7\ s$, i.e. 1 year. This is precisely the value we expected.

The same calculation can be applied to enantiomeric crystals, the d-quartz and the l-quartz for instance. The ratio V_0/W_0 is much greater than in a molecule owing to the crystal framework. The value 70 for this ratio, which is certainly smaller than the actual value, gives an ergodicity time approximately equal to the age of the universe, 20×10^9 yrs. Under these conditions, no spontaneous racemization can be observed for a crystal, which agrees with experiment.

Other approaches of the ergodicity time

Formula (V-4) we have used as a first approach, connects the ergodicity time with the quadratic dispersion of the energy. At first sight, the results seem to be acceptable. Nevertheless, for the molecular systems, such a calculation is obviously wrong. Let us, indeed, consider the case of H_2^+. At the equilibrium distance $\Delta E = 1.2 \ a.u.$, and for an infinite distance, 1.0. These dispersions are thus practically the same, which involves that the corresponding ergodicity times are approximately equal to each other. Now such a conclusion is physically unacceptable. Consequently, formula (V-5) is not applicable and another approach has to be found.

For this purpose, let us consider an harmonic oscillator for which we have found $\tau_e \cong T / \pi$ (V-6) . The average quadratic velocity is given by $\overline{v^2} = E_0 / m$. On the other hand, $\overline{x^2} = E_0 / k$ corresponds to a characteristic length. Let us define the time τ' by

$$\tau'^2 = \frac{\overline{x^2}}{\overline{v^2}} \tag{V-11}$$

Hence,

$$\tau' = \frac{T}{\pi} = \tau_e \tag{V-12}$$

This time τ' corresponds to the average time which is necessary to the particle moving with the average speed $\sqrt{\overline{v^2}}$ for visiting the average domain in which it can be considered as being localized. From which the idea of using this essentially classical calculation in order to determine the order of magnitude of the ergodicity time. We will notice that the value obtained for τ' is the same as that deduced from (V-5).

For an hydrogenic atom of nuclear charge Z, where $\overline{v^2} = \dfrac{Z^2 e^2}{m a_0}$ and a_0 / Z a scale factor, which an average path corresponding to the length of three great circles (the equator + 2 orthogonal meridians) centered on the nucleus and of radius a_0 / Z, we obtain

$$\tau'^2 = \left(\frac{6\pi a_0}{Z}\right)^2 \cdot \frac{m a_0}{Z^2 e^2} = \frac{(6\pi)^2 a_0^2}{Z^4 \alpha^2 c^2} \tag{V-13}$$

α being the fine structure constant $e^2 / \hbar c$ which is equal to $1/137$. The obtained value, $\tau_e \cong 4 \times 10^{-16} s$, is more acceptable than that deduced, from (V-5), namely $2.10^{-17} s$. During this time the electron, indeed, would cover a distance of only 1 Å in the hydrogen atom (the average quadratic velocity of the electron in this atom is of about $c / 100$).

Let us now consider the debatable case of H_2^+. At the equilibrium distance ($r_e \cong 2 a.u.$), the average kinetic energy is equal to $1.1\, e^2 / a_0$, from which $\overline{v^2} = 2.2 \alpha^2 c^2$. If for the characteristic length we choose two times the internuclear distance (the particle goes there and back between the two nuclei), we obtain $\tau' \cong 10^{-16} s$.

Another approach, nevertheless, can be made. Given that, in any case, we will obtain the order of magnitude of the ergodicity time only, we can describe the system by means of a molecular orbital built up upon the atomic orbitals $1s$ of the isolated atoms, a and b. It is well-known that we obtain two states, symmetrical and antisymmetrical respectively with respect to the exchanging of the nuclei

$$\psi_S = \frac{a+b}{\sqrt{2(1+S)}} \quad \text{and} \quad \psi_A = \frac{a-b}{\sqrt{2(1-S)}} \tag{V-14}$$

S being the overlap integral $<ab>$. The energies of these states are respectively the following

$$E_S = \frac{H_{AA} + H_{AB}}{1 + S} \quad \text{and} \quad E_A = \frac{H_{AA} - H_{AB}}{1 - S} \tag{V-15}$$

$$\text{with } \begin{cases} H_{AA} = \left\langle a \left| \hat{T} - \dfrac{1}{r_A} - \dfrac{1}{r_B} + \dfrac{1}{r_{AB}} \right| a \right\rangle \\[2mm] H_{AB} = \left\langle a \left| \hat{T} - \dfrac{1}{r_A} - \dfrac{1}{r_B} + \dfrac{1}{r_{AB}} \right| b \right\rangle \end{cases} \qquad \text{(V-16)}$$

A and B being the two nuclei. ψ_S corresponds to the ground state and ψ_A to an excited level.

Let us now consider the following non-stationary state

$$\Psi = \frac{1}{\sqrt{2}} \left(\psi_S e^{-iE_S t/\hbar} + \psi_A e^{-iE_A t/\hbar} \right) \qquad \text{(V-17)}$$

The electron density

$$\left| \Psi^2 \right| = \frac{1}{2} \left[\psi_S^2 + \psi_A^2 + \psi_S \psi_A \cos(E_A - E_B) t / \hbar \right] \qquad \text{(V-18)}$$

exhibits periodical oscillations between the two nuclei A and B. At $t = 0$, we have $|\Psi|^2 = a^2$, and at $t = \pi \hbar / (E_A - E_B)$, $|\Psi|^2 = b^2$. The frequency of these oscillations is $\nu = (E_A - E_B) / h$. The conventional quantum mechanics interprets this result by saying that the electron oscillates from one nucleus to the other with the frequency ν. The argument is the same as that used for the enantiomers. Nevertheless, as we have said, such an interpretation is not valid for the ground state. Various authors, Eyring *et al.* [9] in particular, have pointed out the fact that this result "should not be taken too literally". However, we can use this result for our discussion. Indeed, if we start from a density localized around the nucleus A, owing to the electron radiation which is not compensated by the universe field because the system is not in equilibrium, we will have an oscillation of the electron density with a damping which will cause the spreading of the density : $|\Psi|^2 \rightarrow \psi_S^2$. As a first approximation, we can assume that, in spite of the damping, the frequency of the oscillations remains the same. In consideration of which, the ratio $h / \pi (E_A - E_B) = \tau''$ can be considered as the average time which is necessary for the electron to visit the whole molecular domain. The factor π arises from Eq (V-6).

At the equilibrium distance, the difference in energy $(E_A - E_B)$ is of about 0.3 e^2 / a_0 . Consequently $\tau'' \cong 2 \times 10^{-16} s$. We find again the same order of magnitude as that previously obtained.

This kind of approach allows an interesting application. It is, indeed, obvious that the ergodicity time in any molecule does depend on the internuclear distances. Greater the latter, larger the time taken by the electron for visiting the whole molecular domain, and, consequently, for realizing the quantum density. On the other hand, when the nuclei are sufficiently removed from one another, the quantum description loses its meaning because, physically, from a certain distance the atoms cannot be considered as being bound, otherwise any nucleus and any electron which would have successively belonged to a huge number of systems since billions of years, should keep bonds with all the systems or, at least, the memory of systems which do not exist at the present time.

In the case of H_2^+, it is easy to verify that τ'' (thus τ_e) increases very quickly as the internuclear distance R increases. If we admit that H_{AB} varies as the overlap integral S between the two orbitals of the atoms, we obtain the values reported in the following Table, which are, physically, more satisfying than those deduced from Eq (V-6).

R a.u.	S	$\tau'' \cong \tau_e$
2.5	0.3	$2 \times 10^{-16} s$
10	2×10^{-3}	$3 \times 10^{-14} s$
20	3×10^{-7}	$2 \times 10^{-10} s$
40	2×10^{-15}	0.03 s
50	2×10^{-19}	5 mn
60	1×10^{-23}	3 months
80	3×10^{-30}	10^4 yrs

These values show that the quantum description with the strong correlation it involves between the components of the system, is not valid for distances greater than twenty Ångströms, the time necessary for the quantum density to be reached, becoming macroscopic.

As a last example, we will examine the case of a particle of mass m constrained to move freely in a rectangular box with edges of length a, b, c $(a > b > c)$: $0 \leq x \leq a$; $0 \leq y \leq b$; $0 \leq z \leq c$. The wave-function $\psi(x,y,z)$ is the product of three functions $X(x)$, $Y(y)$, $Z(z)$ corresponding to the three dimensions respectively and the energy is the sum of the three energies E_x, E_y, E_z associated with these functions. For the motion along x, the energy of the ground state is the following

$$E_x = \frac{h^2}{8ma^2} \qquad \text{(V-19)}$$

This energy is kinetic in origin only, so that the average quadratic velocity is

$$\sqrt{v^2} = \frac{h}{2ma} \qquad \text{(V-20)}$$

For the characteristic length it is logical to adopt a. From which the corresponding ergodicity time

$$\tau_e \cong \frac{2ma^2}{h} \qquad \text{(V-21)}$$

i.e. for an electron

$$\tau_e(s) \cong a^2 (cm^2) \qquad \text{(V-22)}$$

For the lengths whose order of magnitude corresponds to the atomic size, i.e. $a \cong 1 a.u.$, the ergodicity times are of about $10^{-16} s$. Thus we find the same order of magnitude as the one obtained by the preceding calculations. Nevertheless, the increasing of this time versus a is much slower than for the systems in which the electron is submitted to a non zero potential, e. g., for $a = 1\mu = 10^{-4} cm$, $\tau_e \cong 10^{-8} s$.

Likewise, we obtain the ergodicity times for the two other dimensions y and z. These times are proportional to b^2 and c^2 respectively. Now the ergodicity will be achieved for the system when it will be reached for each of the three dimensions, i.e., in fact, for the greatest of the dimensions. Consequently the ergodicity time for the particle in the box under consideration is given by (V-22) since $a > b > c$.

Going back to the measurement axiom

The remark made at the beginning of this chapter concerning the fact that any measurement demands a finite time, could lead us to think that the general quantum framework of the orthodox quantum mechanics can be kept, provided that, in the measurement axiom, we make precise that it is a question of purely theoretical instantaneous measurements, whereas in the reality the observation time is, in the general case, such as the limiting average value $\bar{g} = \sum g_i |C_i|^2$ is reached. However, such a interpretation does not agree with our model owing to the fact that, in the latter, all the properties fluctuate in course of time. The measurement axiom has to be replaced by a weaker one, namely : *Any sufficiently long mesurement gives the time-average value of the property under consideration.* We will see the importance of this statement in connection with the spin of the particles (Chapter IX) .

References

[1] W.H. Zurek, *Phys. Rev.* **D 26** (1982) 1862 ; R. Griffiths, *J. Stat. Phys.* **36** (1984) 219 ; R. Omnès, *Rev. Mod. Phys.* **64** (1992) 339 and *The interpretation of Quantum Mechanics* (Princeton University Press, Princeton NJ) 1994.

[2] M. Bunge, *Int. J. Quantum Chemistry* **XII**, suppl. **1** (1977) 1.

[3] S. Haroche, J.M. Raimond, M. Brune, *Phys. Rev.* **77** (1996) 4887.

[4] R. Omnès, *Phys. Rev.* **A 56** (1997) 3383.

[5] Chr. K. Jörgensen, *Modern aspects of ligand theory* (North-Holland, Amsterdam) 1971.

[6] H.P. Lemaire, R.L. Livingston, *J. Chem. Phys.* **18** (1950) 569 ; W. Edgell, D.G. Weiblen, *ibidem* **18** (1950) 571.

[7] A. Julg, *Folia Chimica Theoretica Latina* **XIX** (1991) 73.

[8] L. Landau , E. Lifchitz, *Physique Théorique : III Mécanique Quantique* (Mir; Moscou) 1966, p. 182.

[9] H. Eyring, J. Walter, G.E. Kimball, *Quantum Chemistry* (Wiley, New York) 1944, p. 199.

VI

The molecular structure

The Born-Oppenheimer approximation

The chemist considers that the molecules which he prepares and he studies, possess well-defined forms, just like the bodies which surround him. These forms, as we have seen in Chapter IV, are determined by the average position of the nuclei, the small amplitude oscillations of the nuclei around their respective equilibrium position making the structure of the molecules flexible, without forbidding all that to ascribe a form to them. The electrons constitute the cement which insures the cohesion between the nuclei.

The quasi-totality of the calculations performed by quantum mechanics is based on this essentially classical image. The nuclei are assumed as being at rest, located at their equilibrium position, so that the wave function obtained by integrating the Schrödinger equation is purely electronic. That is the Born-Oppenheimer approximation. Such a procedure allows to reach the most important properties of the molecules, as the dipole moment, the absorption spectrum as well as interesting data about their reaction capacity. In order to obtain a more complete description of the molecule, its vibration spectrum in particular, Born and Oppenheimer [1] use the following wave-function

$$\Psi(e,n) = \psi_{\text{elec.}}(e,n_0) . \psi_{\text{vibr.}}(n-n_0) \qquad \text{(VI-1)}$$

where $\psi_{\text{elec.}}(e,n_0)$ is the purely electronic function determined by assuming that the nuclei are located at their equilibrium position n_0, and

$\psi_{vibr.}(n - n_0)$ a purely vibrational nuclear function, n being the instantaneous position of the nuclei.

Within the framework of our model, it is easy to justify this approximation. We have seen, indeed, that the value of the electron ergodicity time $(10^{-16} s)$ is very weak with respect to that corresponding to the nuclear oscillations $(\approx 10^{-12} s)$. It follows that, for any position of the nuclei, we can consider that the electrons have plenty of time to realize the density ascribed by quantum mechanics, so that, every thing occurs as if the nuclei were submitted to the time-independent average potential created by the whole of the electrons, which allows to define the purely vibrational function $\psi_{vibr.}(n - n_0)$.

The Born-Oppenheimer approximation therefore assumes implicitly that the electronic ergodicity time is very different from the vibration periods. Consequently, this approximation will be not valid if these two times are of the same order of magnitude, as in the ion H_2^+ for internuclear distances greater than 20 $a.u.$ $(\approx 10\text{Å})$. The electron ergodicity time is, indeed, greater than $10^{-7} s$, i.e. it reaches the order of magnitude of the vibration period. In others words, the Born-Oppenheimer approximation is meaningful only for the molecules around their equilibrium distance.

On the other hand, the rotation periods are, in the general case, widely greater than the vibration ones, so that the molecule behaves as a rigid body. Function (VI-1) will be multiplied by a purely rotational function of the Euler angles (η, ξ, χ) which define the orientation of the molecule, assumed as being a rigid body

$$\Omega = \psi_{elec.}(e, n_0) \cdot \psi_{vibr.}(n - n_0) \cdot \psi_{rot.}(\eta, \xi, \chi) \qquad \text{(VI-2)}$$

The Woolley paradox

The Born-Oppenheimer approximation is implicitly based on the idea that the molecules possess a form. The minimization of the electron energy with respect to the coordinates of the nuclei which allows to obtain the equilibrium geometry, contains this concept. Now the total hamiltonian of the system, i.e. including the coordinates of the nuclei and those of the electrons, is invariant under any rotation around the center of gravity of the system. From this it results that the total wave-function Ω (VI-2) exhibits the same property. In other words, the corresponding density is spherical, which finally significates that *all molecules are spherical* . That is the Woolley paradox [2] . The conclusion is more especially serious as, at the limit, all the bodies which surround us would be spherical ! Not only, the molecular concept is not found again by quantum mechanics, but also it is even categorically at variance with the latter.

In fact, the paradox is not real . No paradox can exist in Science, because that would be the denial of the latter. The key of the problem is in the ratio ergodicity time of the observed phenomenon over the observation time. If quantum mechanics does effectively give values corresponding to observations longer than the ergodicity time, given that the molecules are unceasingly rotatoring (III-29) , it is obvious that the nuclei density is spherical. Consequently, an observer which would photograph the system using a sufficiently long exposure, would obtain an image which he will interpret as arising from a spherical body. But, if he reduces the exposure in a suitable manner, he will see the image of a body exhibiting a well characterized form appear which, besides, will not necessarily be the same for successive views. In passing, we will notice that the paradox cannot be removed within the orthodox quantum framework, the latter leading to a rotation energy equal to zero.

Case of isomers

Two molecules are called *isomers* when they have the same empirical formula, e.g. methyl-acetylene and allene C_3H_4, or naphtalene (two fused hexagonal cycles) and azulene (two fused cycles, the one pentagonal and the other heptagonal) $C_{10}H_8$.

The interpretation of the Woolley paradox we have just proposed, allows to throw a new light on the problem of these molecules. Indeed, the Hamiltonian operators corresponding to these pairs of molecules are different within the Born-Oppenheimer approximation, but identical if we introduce the coordinates of the nuclei. In the same way as quantum mechanics, neglecting violating parity interactions, obtains a unique function to describe two enantiomers. In other words, this theory should yield the same function for naphtalene and azulene ! The chemist cannot, of course, accept such a conclusion, because naphtalene and azulene are well-defined molecules, exhibiting specific physical and chemical properties. Besides nobody has seen naphtalene to be spontaneously changed into azulene, and conversely.

The problem of two isomers whose energies are necessarily different, corresponds formally to that of a particle in a dissymmetrical double-well potential. In the case of several isomers corresponding to the same formula, the potential exhibits a greater number of wells, but the problem remains the same, so that we will consider the case of a double-well only. The arsines have shown to us that over a range of time weaker than the ergodicity time, each of the isomers can be separately observed, but that after one year, in the average, the optical activity is equal to zero, which corresponds accurately to which we will call the "quantal form". For two isomers, the situation is the same. An observation over an extremely long time, a thousand years perhaps (or more) , would give the average "quantal form". But, at our time scale, we have the possibility to study each of the forms corresponding to the various wells separately and to observe chemical reactions. Still here, that is a matter of time.

The chemical bond

Let us consider a time interval sufficiently brief in order that the molecules can be said of as having a classical sense, i.e. with nuclei practically at rest, located within a given potential well, and let us examine the problem of the electrons. Since the Lewis works [3] , the chemists identify the dashes which they draw between the nuclei of the atoms, to localized electron pairs, for instance for methane CH_4

$$
\begin{array}{ccc}
H & & H \\
 & \diagdown C \diagup & \\
 & \diagup \diagdown & \\
H & & H
\end{array}
$$

Now quantum mechanics ascribes the same average properties to all the electrons of the system, in particular, the same average position.

An usual manner to remove the contradiction between the two models consists to say that the electrons are localized during a certain time in the various pairs respectively, then they exchange with one another so as to visit all the sites after a sufficiently long time. Such a manner to see the phenomena allows to explain the fundamentally local character of the chemical reactions. At the begining of the reaction, a given electron pair is attacked without the other pairs being affected. In fact, this explanation is worth examining a little more closely.

In this view, let us consider a system with an even number of electrons. The latter can be described by a Slater determinant , built up upon functions φ_i, called atomic or molecular *orbitals* according as the system under consideration is an atom or a molecule. These functions are doubly utilized, associated successively with the spin functions α and β

$$\psi(1,2,...n) = \frac{1}{\sqrt{n!}} \begin{vmatrix} \varphi_1(1)\alpha(1) & \varphi_1(1)\beta(1) & \cdots & \varphi_{n/2}(1)\beta(1) \\ \varphi_1(2)\alpha(2) & \varphi_1(2)\beta(2) & \cdots & \varphi_{n/2}(2)\beta(2) \\ \cdots\cdots & \cdots\cdots & \cdots & \cdots\cdots \\ \varphi_1(n)\alpha(n) & \cdots\cdots & \cdots & \cdots\cdots \end{vmatrix}$$

(VI-3)

As it is easy to verify, all the electrons play the same role and exhibit the same average properties. Indeed, for any property G

$$\overline{G(1)} = \int \psi*(1,2,...n)\hat{G}(1)\psi(1,2,...n)dv$$
$$= \frac{2}{n}\left[<\varphi_1\hat{G}\varphi_1> + <\varphi_2\hat{G}\varphi_2> +...+ <\varphi_{n/2}\hat{G}\varphi_{n/2}>\right] \quad \text{(VI-4)}$$

From which it results

$$\overline{G(1)} = \overline{G(2)} =...= \overline{G(n)}$$

(VI-5)

This result is often interpreted by saying that the electrons of the system are *indistinguishable* owing to the fact that they exhibit the same properties. In fact, we see that this description is a artifact arising from the structure of Quantum Mechanics which considers the system as a whole. In reality, well-determined properties can be assigned at every time to the various electrons.

As a first example, let us consider a beryllium atom. This atom possesses four electrons. For the atomic functions φ_i, we will use hydrogenic $1s$ and $2s$ orbitals, with suitable nuclear charges. The analysis of the total radial density $(4\pi r^2 \psi^2)$ leads to the distribution indicated on the figure [4]

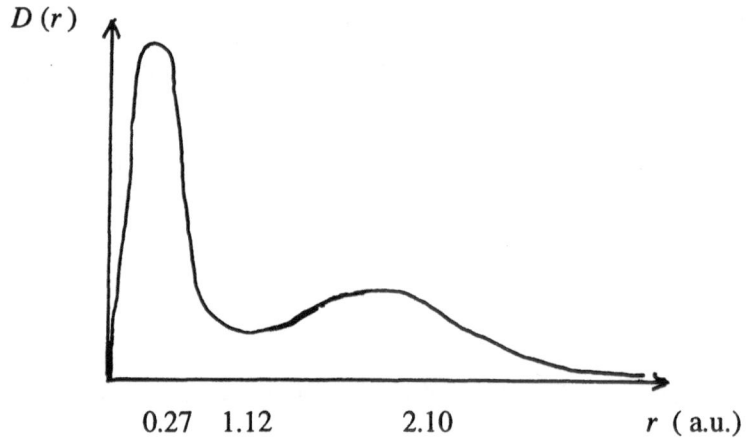

Radial electron density in Be atom.

This distribution can be interpreted as arising from the superposition of two shells whose respective densities exhibit maximums at 0.27 and 2.10 a.u. respectively. By integrating this density, it is shown that the sphere of 1.12 a.u. radius contains two electrons in the average, thus, that the two other electrons are located outside this sphere.

In order to make the description more precise, we will compute the relative fluctuation of the electron population of the various domains of space. The *relative fluctuation* is equal to the square root of the variance $\overline{N}$ of the population, divided by the average value $\overline{N}$ of the latter

$$f = \frac{\sqrt{\overline{\overline{N}}}}{\overline{N}} \qquad\qquad \text{(VI-6)}$$

Let us recall that $\left(\overline{\overline{N}}\right)^2 = \overline{N^2} - \left(\overline{N}\right)^2$. A simple calculation shows that the relative fluctuation of the electron population inside a sphere Σ centered at the nucleus, is minimal for a radius of about 1.1 a.u. , i.e. for the sphere which corresponds to the minimum of the radial density $D\,(r)$. The value of this relative fluctuation is equal to 0.15. Consequently, the fluctuation corresponding to the domain exterior to this sphere is also equal to 0.15. This result means that, in the average, two electrons are localized inside the sphere of 1.1 a.u. radius, and two outside the latter, the electrons being unceasingly exchanging with one another. We join the conception of the chemist again, justifying the inner electron concept and explaining their chemical inertia. Whereas the four electrons of the beryllium atom are equivalent, two only are able to be implicated in the reactions. That is the reason for which one obtains BeH_2 and not BeH_4. Likewise, one would see that in the carbon atom which possesses 6 electrons, 4 only are available for reactions, 2 electrons being localized within what the chemist calls the "inner shell".

The problem of molecules is more complex. In order to built up the molecular orbital φ_i , one uses linear combinations of the atomic orbitals of the isolated atoms χ_r

$$\varphi_i = \sum_r c_{ir}\chi_r \qquad\qquad (VI\text{-}7)$$

The densities $|\varphi_i|^2$ are generally delocalized over the whole molecule so that an analysis analogous to that performed for Be is not possible.

A first approach consists to determine regions of space, called *loges* , joined together but without overlapping, such as within each loge the average number of electrons is equal to two. The partitioning of space we obtain, is in fact very artificial and it is always more or less based on the bond concept we will find again ! A more interesting concept is that of loge with a *minimal fluctuation* [5] . The problem consists in defining regions of space, joined together and without overlapping, where the average number $\overline{N}$ of electrons located inside each domain, exhibits a minimal fluctuation. The calculations are rather tedious, but interesting results have been obtained for LiH, BeH_2, CH_4. From the total density, one finds again the concept of electron pairs localized on an average between pairs of nuclei. Unfortunately, such loges do not exist in NH_3 , OH_2 or N_2, which shows not only that quantum mechanics cannot justify the chemical bond concept, but also that this notion is incompatible with this theory. The chemist, of course, cannot accept such a conclusion. He cannot believe that such a concept which has allowed the prodigious rise of chemistry for the last hundred fifty years, does not correspond to any physical reality, even if quantum mechanics cannot reveal it.

The orbital domains

There is another possibility, however, which supports our general interpretation of quantum mechanics.

First, we will remark that the monodeterminantal structure of the wave-function (VI-3) allows to replace the set of the doubly occupied functions φ_i by a set of functions φ_i', also doubly occupied, deduced from the functions φ_i by means of any unitary transformation. The total function (VI-3) is, indeed, invariant under such a transformation, so that the global quantum description of the system is unchanged. Taking advantage of this indetermination of the molecular orbitals, we can seek the set of the functions φ_i' which would present a maximal degree of localization. For instance, we can demand that the sum of the Coulomb repulsions between the electrons for each of the new partial densities is maximal [6] .

Such an operation leads to densities $|\varphi_i'|^2$ which exhibit all the characteristics of the Lewis electron pairs [7] . We find again the inner shells (as for Be) and functions strongly localized between pairs of nuclei, corresponding to the chemical bonds, as well as functions concentrated at the vicinity of certain nuclei, but excentred with respect to the corresponding nuclei, and which can be identified with the classical lone pairs (e.g. in NH_3).

Practically, the functions φ_i' which are localized between the nuclei reduce as follows

$$\varphi'_{AB} = \sum_{a \in A} c_{Aa} \chi_{Aa} + \sum_{b \in B} c_{Bb} \chi_{Bb} \tag{VI-8}$$

χ_{Aa} and χ_{Bb} being orbitals carried by the nuclei A and B respectively. In fact, it remains a very weak contribution arising from orbitals carried by the other nuclei in the functions φ'_{AB}. The localization is not complete. But the effect can be neglected. By regrouping the orbitals χ_{Aa} into a unique function t_A, and the orbitals χ_{Bb} into a function t_B , we obtain

$$\varphi'_{AB} = C_A t_A + C_B t_B \tag{VI-9}$$

The functions t_A and t_B which are linear combinations of the orbitals χ_{Aa} and χ_{Bb} respectively, appear as being a new basis to built up the functions φ'_{AB} . They are called *hybrid orbitals*. The densities $|\varphi'_{AB}|^2$ are strongly concentrated between the nuclei A and B. Likewise for the densities t_A^2 and t_B^2 corresponding to the hybrids. These orbitals exhibit a strong dissymmetry. The center of gravity of each of them is, indeed, removed towards the nucleus which carries the other. For instance, in methane CH_4, the $2s$ orbital and the three $2p$ orbitals of the carbon atom are replaced by the four following hybrid orbitals

$$\begin{cases} t_1 = \dfrac{1}{2}\left[(2s) + (2p_x) + (2p_y) + (2p_z)\right] \\ t_2 = \dfrac{1}{2}\left[(2s) + (2p_x) - (2p_y) - (2p_z)\right] \\ t_3 = \dfrac{1}{2}\left[(2s) - (2p_x) + (2p_y) - (2p_z)\right] \\ t_4 = \dfrac{1}{2}\left[(2s) - (2p_x) - (2p_y) + (2p_z)\right] \end{cases} \tag{VI-9}$$

which respectively point towards the hydrogen atoms located at the apices of a regular tetrahedron whose center is the nucleus of the carbon atom. The centers of gravity of the corresponding electron densities are respectively located between the nucleus C and the corresponding H atom.

p (s,p)

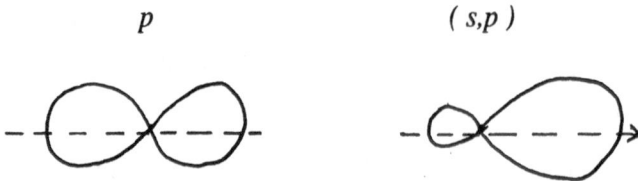

Pictorial representation of the electron density for a p -pure orbital and a (s,p) - hybrid orbital.

Consequently, we can consider the molecule as formally resulting from molecular orbitals built up upon pairs of hybrids pointing at one another. This interpretation, of course, agrees to many chemists which see the foundation of stereochemistry in it. But the phenomena are more complex because the model does not correspond to a physical localization of the electrons. Nevertheless, we are entitled to think that in order to find the classical chemical bond concept again, such hybrid orbitals must play a preferential role.

Let us analyze the total density $|\psi|^2$ in terms of contributions arising from the various pairs of orbitals (t_A, t_B) . If the atoms A and B are "chemically bonded", the relative fluctuation of the population of the domain defined by the set of the two orbitals t_A and t_B , we will call *orbital domain* , is always weaker than 0.10 . On the contrary, for domains built up upon two orbitals which do not point at one another, the fluctuation is greater than 0.5 . For instance, in methane, for the domain which corresponds to a bond C-H (built up upon a hybrid orbital issued from the carbon atom and the atomic orbital $1s$ of the corresponding hydrogen atom), the relative fluctuation is equal to 0.06 , while for a pair corresponding to two hydrogen atoms, $F=0.5$. This rule is absolutely general [7] .

Consequently, the chemical bond can be effectively interpreted as corresponding to orbital domains of weak relative fluctuation around an average value of the charge, in general of about 2 electrons, which involves that the electrons of these various domains are randomly exchanging between one another at more or less wide intervals, thus insuring their complete delocalization on a sufficient long time. On the other hand, we will notice that this analysis gives a physical meaning to the hybridization concept which, at first sight, could have been considered as a pure mathematical artifact [8] .

In a conjugated molecule, as benzene, a high relative fluctuation occurs for any domain built up upon two orbitals $2p_z$ carried by two adjacent carbon atoms, which corresponds to an important delocalization for the electrons. On the contrary, if we consider the orbitals π of the conjugated system as a whole, at the second order with respect to the overlap integrals, we obtain a relative fluctuation equal to zero, which justifies the classical distinction between the strongly localized underlying σ-electrons and the very mobile π-electrons. Contrary to what occurs in a saturated molecule, a conjugated system cannot, indeed, be attacked at a given point without being entirely affected.

An important point has to be emphasized. The orbital domains we have introduced, do not correspond to well-determined area of space, as a chemist would have expected. Even if, conventionally, we assign a "volume" defined by an isodensity surface in which, in the average, a given fraction of the electron charge - say 95% - is confined, to the orbitals t , the volumes corresponding to the various bonds issued from a given nucleus - e.g. the four C-H in methane - , are overlapping so that the chemical bond cannot be explained within the 3-D geometrical space. It is necessary to consider a more extended space.

The first idea is to represent these domains within the phase space (q, p). The "volumes" of the various orbital domains which are overlapping within the geometrical space $(q$) are, indeed, disjoined within the phase space. Moreover, in this way, the dynamical character of the chemical bond is automatically introduced. Unfortunately, it is well-known that the phase space is not adapted to quantum mechanics, the density within this space being not always positive or equal to zero.

We will subsequently see (Chap. XI) how our model allows to answer to the problem of the very nature of the orbital domains.

Whatever that may be, the introducing of orbital domains allows to reconcile the localized bond concept with the completely delocalized quantum description within the framework of a quasi-ergodic interpretation of Quantium Mechanics, the latter giving the average description of the molecule over a sufficiently long time. The meaning of this result which goes beyond the simple framework of the problem of the chemical bond constitutes a new argument for our interpretation.

Case of crystals

Although crystals have to be considered as giant molecules [9] , their theoretical study presents a conceptual difficulty owing to their macroscopic size. We have indeed shown in Chapter V that when the interatomic distances are increasing, the ergodicity times very quickly become so great that the quantum description loses its meaning. In fact, the difficulty can be got round through an artifice based precisely on the notion of ergodicity.

In first, let us recall that the crystals can be classified into three categories:

i. the *molecular* crystals formed by a stacking of small sized molecules (inert gases, nitrogen, oxygen, water, carbon dioxide and the greatest number of organic compounds) .

ii. the *macromolecular* crystals, consisting in an infinite tridimentional lattice of bonds which are more or less polar and localized between nuclei (diamond, oxides, sulfurs, silicates, halides,...) .

iii. the *metals* in which certain electrons are completely delocalized over all the bulk.

In the first case (i) , we can consider that the ergodicity is reached for every small molecule of the crystal, so that the problem reduces to that of *one* of these molecules within the multipolar average field created by the others, without being necessary to write the wave-function corresponding to the whole crystal (The classical chemistry refers to dipolar interactions and to van der Waals forces) . The difference in energy between the one of the molecules inside the bulk and the same but isolated molecule gives the cohesion energy.

For the macromolecular crystals (i i) , we can admit that the ergodicity is reached for the electron pairs of the various orbital domains owing to the small length of the corresponding bonds, so that the problem reduces to that of *one* electron pair inside a given orbital domain within the average field created by the other pairs assumed as being localized in their respective orbital domain. In this way, we can define a wave-function for each electron pair, which allows to determine the properties of the corresponding bonds and, through the latter, the ones of the crystal, its ionicity in particular [10] . Here still, it is not really useful to construct the wave-function associated with the bulk for the knowledge of the properties of the crystal, which clearly shows the artificial nature of the wave-function. In passing, we will notice that the problem is the same for macromolecules, e.g. the nucleic acids, and for glasses which differ from crystals only in their disorder.

Concerning the metals (i i i) the situation is different. For instance, in a univalent metal (M =Li, Na, K) , in the mean, *one* electron is localized in each site M with an unceasing exchange between the sites. Indeed, the relative fluctuation of the charge of each atom (0.71 [4]) involves a strong mobility of the electrons. Now we have seen in the preceding chapter that in this case the ergodicity times remain very weak even for macroscopic sizes ($10^{-8}s$ for a crystal of 1μ) , which gives a meaning to the quantum description.

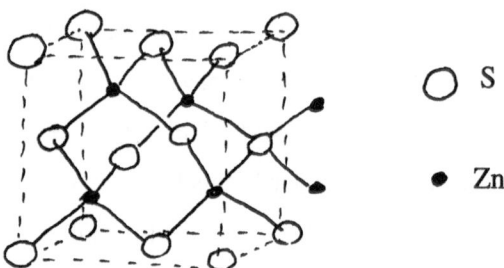

Examples of macromolecular crystals:

Zinc-blende ZnS : Each Zn atom is linked to four S atoms. The bonds are identical and partially ionic (The net charges of Zn and S atoms are equal to +1.5 and -1.5 respectively [9]) .

Diamond : The atoms are of the same nature (C) . The structure is the same as that of ZnS. The bonds are not polar (covalent bond) .

Does Quantum Mechanics apply to the universe considered as a giant molecule ?

The question can appear as being rather absurd. In fact, it is worth being set owing to the fact that the wave-functions and consequently the electron densities are rigorously vanishing only at the infinity. According to the orthodox Quantum Mechanics, an electron localized at present time on Earth can be sometime or other on Saturn or in a far galaxy ! In other words the situation would be comparable with the one encountered for molecules in which the orbitals carried by the various nuclei are overlapping and consequently combine with one another to ensure the stability of the system.

In fact, the situation must be considered in another view. If nothing forbids an electron to visit the whole universe (for instance, through successive exchanges between the molecules which constitute the latter) and if the universe effectively forms a whole, it cannot be described by Quantum Mechanics as a giant molecule, for two reasons at least :

First, more still than for crystals, the huge size of the universe (with respect to the atomic scale) precludes the ergodicity to be reached so that quantum description has no meaning for us.

Secondly, if Quantum Mechanics is a means to take the interactions between the system under consideration and the systems which surround it, into account, this mechanics cannot be applied owing to the fact that, by definition, nothing does exist out the universe or, at the very least, no exhange is possible with matter which would be located outside our universe.

The universe is an entity which has to be considered as a whole within a classical framework, while Quantum Mechanics is reserved to systems of small size belonging precisely to the universe.

References

[1] M. Born, J.R. Oppenheimer, *Ann. der Physik* **84** (1927) 457.

[2] R.G. Woolley, *Adv. Phys.* **25** (1976) 27 ; *J. Amer. Chem. Soc.* **100** (1978) 1073.

[3] G.N. Lewis, *J. Amer. Chem. Soc.* **38** (1916) 762.

[4] A. Julg, *La Liaison Chimique* (Presses Univer. France, Paris) 1980.

[5] For inst. see : R. Daudel, H. Brion, S. Odiot, *J. Chem. Phys.* **23** (1955) 2080 ;

R. Daudel, *Fundamentals of theoretical Chemistry* (Pergamon Press, Oxford)

1968 , and references quoted in [7] .

[6] K. Ruedenberg, in *Modern Quantum Chemitry, Istanbul Lrcture,* O. Sinanoglu ed.

(Academic Press) 1965, Part I, p. 85.

[7] A. Julg, P. Julg, *Int. J. Quantum Chem.* **13** (1978) 483 ; A. Julg, *ibidem* **36**

(1984) 709.

[8] A. Julg, *J. Mol. Structure (Theoc hem)* **169** (1988) 125.

[9] A. Julg, *Crystals as giant molecules* , Lecture Notes in Chemistry **9** (Springer-Verlag , Berlin) 1978.

[10] A. Julg, F. Marinelli, A. Pellégatti , *Intern. Journal of Quantum Chemistry* , **XIV** (1978) 181 ; A. Julg, A. Pellegatti, F. Marinelli, *Israël Journ. of Chemistry* , **19** (1980) 260.

VII

A mathematical approach

The model

The few examples we have studied clearly show the advantages of the physical interpretation envisaged to account, qualitatively at least, for the chief experimental features and for a number of situations which are considered as being typically quantal, i.e. not classically interpretable by nature. In this chapter, we purpose to built up an essentially classical mathematical model, able to justify the proposed interpretation [1].

In this view, we will consider a particle of mass m and of charge q, submitted :

i. to a deterministic force $\mathbf{F(R)}$,

ii. to the electromagnetic damping force $-\mathbf{f}$,

iii. to an electromagnetic field $(\mathbf{E}, \mathbf{H})$ designed to simulate the universe field, i.e. the field created by the whole of the other systems of the universe.

The general equation which governs the motion of the particle is the following

$$-\mathbf{f} + \dot{\mathbf{p}} = \mathbf{F(R)} + q\mathbf{E} + q\frac{\dot{\mathbf{R}}}{c} \wedge \mathbf{H} \qquad \text{(VII-1)}$$

where $\mathbf{p}$ is the momentum of the particle.

According to the ideas developed in Chapter II, we will assume that the field of the universe is constituted by the succession of non-correlated very sudden and very brief trains of oscillations (*bursts*), travelling at the speed of light, each of them along its own direction, the oscillations being completely damped before the beginning of the following burst, so that, in the time, two successive bursts do not overlap, which allows us to define the life of each burst. Owing to the average isotropy of the universe, the direction of propagation of the various bursts, in the mean, must be oriented in an isotropic manner. Moreover, the variations of the field must exhibit an ergodic behavior (Stability of the averages of the various properties over a sufficiently long time intervall) .

Concerning the damping force, its expression, in the general case, is very complex. The reader will find it in Ref. [2] . Nevertheless, if we neglect the terms in $\beta^2 = v^2 / c^2$ arising from relativistic corrections, we obtain the following expression which is easier to handle

$$f = \tau q \left(\dot{E} + \frac{\dot{R}}{c} \wedge \dot{H} \right) + \frac{\tau q^2}{mc} \left[E \wedge H + \left(\frac{R}{c} \wedge H \right) \wedge H + E \wedge \left(E \wedge \frac{R}{c} \right) \right]$$

with $\tau = \dfrac{2q^2}{3mc^3}$ $\qquad\qquad$ (VII-2)

which appears as a characteristic time attached to the particle (for electron, $\tau \approx 6 \times 10^{-24} s$).

In fact, the last term in the brackets whose absolute value is weaker than $\tau q^2 E^2 / mc$, can be neglected with respect to the electronic force $q E$ (VII-1) provided that $|E| < 10^{16} ues$. This condition can be considered as being *de facto* realized owing to the fact that from such a value of the field, the latter would materialize into an electron-positron pair.

On the other hand, to explicit the term in τ of the damping force, we will use the equation of the motion (VII-1) by putting $\tau=0$, i.e.

$$m\ddot{R} = F + qE + q\frac{\dot{R}}{c} \wedge H \qquad\qquad (VII-3)$$

Then, by substituting in (VII-1)

$$f = \tau m \dddot{R} - \tau \dot{F} - \frac{\tau q F}{mc} \qquad\qquad (VII-4)$$

Now, the relative variation of the deterministic force $(\dot{F} / F)$ is negligible with respect to $1/\tau$, so that the term $\tau \dot{F}$ can be neglected with respect to the damping force. Likewise for the last term of (VII-4) which, in the absolute value, is weaker than $\tau q|\mathbf{FE}|/mc$ (for an electromagnetic radiation, the magnetic field exhibits the same value as the electric field) . This term is negligible if $|E| < 10^{16}$ ues.

Consequently, by neglecting the terms of the second order in β , the equation of the motion becomes

$$-\tau\dddot{\mathbf{R}} + \ddot{\mathbf{R}} = \frac{\mathbf{F(R)}}{m} + \frac{q}{m}\mathbf{E_R}(t) + \frac{q}{m}\frac{\dot{\mathbf{R}}}{c} \wedge \mathbf{H}(t) \qquad (VII\text{-}5)$$

where $\mathbf{E_R}(t)$ is the value of the electric field at the time t , at the point $\mathbf{R}$. Owing to the fact we neglect the second order in β , the propagation effect of the magnetic field does not appear, from which the absence of index for the latter. In passing, we will notice that the expression of the damping force $(\tau m\dddot{\mathbf{R}})$ is identical with that deduced from elementary non relativistic considerations, i.e. by neglecting the terms in β [3] .

Study of the motion

The structure of Eq (VII-5) allows us to separate the effects of the three acting forces (deterministic, electric and magnetic) from one another. In a first step, we will not take the magnetic force into account, i.e. we will neglect the terms in β . Eq (VII-5) reduces as follows

$$-\tau\dddot{\mathbf{r}} + \ddot{\mathbf{r}} = \frac{\mathbf{F(r)}}{m} + \frac{q}{m}\mathbf{E}(t) \qquad (VII\text{-}6)$$

(The electric field depends on the time only, given that the terms in β are neglected. On the other hand, the reader will notice the substitution of the coordinate $\mathbf{R}$, corresponding to the complete equation (VII-5) , by $\mathbf{r}$)

Let us assume that at $t=0$ the particle is located at the point $\mathbf{M_0(r_0)}$ with the speed $\dot{\mathbf{r}}_0$ and the acceleration $\ddot{\mathbf{r}}_0$. In the absence of electric field, from this point, the particle would describe an arc $\mathbf{C}$, defined by the following equation

$$-\tau\dddot{\mathbf{C}} + \ddot{\mathbf{C}} = \frac{\mathbf{F(r_0 + C)}}{m} \qquad (VII\text{-}7)$$

(In fact, the term $\tau\ddot{\mathbf{C}}$ is completely negligible)

If we take the universe field (here reduced to the sole electric part) into account, this equation becomes

$$-\tau(\dddot{\mathbf{C}}+\dddot{\mathbf{T}})+(\ddot{\mathbf{C}}+\ddot{\mathbf{T}})=\frac{1}{m}F(\mathbf{r_0}+\mathbf{C}+\mathbf{T})+\frac{q}{m}\mathbf{E} \qquad \text{(VII-8)}$$

with $\mathbf{r}=\mathbf{r_0}+\mathbf{C}+\mathbf{T}\,.\,\mathbf{T}$ and its derivatives are equal to zero at $t\;=\;0$. Consequently, by difference, to the first order in $\mathbf{T}$, owing to the fact we can neglect the deterministic contribution with respect to that arising from the very sudden variation of the universe field, we obtain

$$-\tau\ddot{\mathbf{T}}+\ddot{\mathbf{T}}=\frac{F'(\mathbf{r_0})}{m}\mathbf{T}+\frac{q}{m}\mathbf{E}\approx\frac{q}{m}\mathbf{E}=-\frac{q}{mc}\dot{\mathbf{A}} \qquad \text{(VII-9)}$$

A being the vector potential associated with the universe field.

From which, by integrating

$$-\tau\ddot{\mathbf{T}}+\dot{\mathbf{T}}=-\frac{q}{mc}(\mathbf{A}-\mathbf{A_0}) \qquad \text{(VII-10)}$$

i.e. $\quad \dot{\mathbf{T}}(t)=\frac{q}{mc\tau}e^{t/\tau}\int_0^t(\mathbf{A}-\mathbf{A_0})e^{-t'/\tau}dt' \qquad \text{(VII-11)}$

$$\mathbf{T}(t)=\frac{q}{mc}\left[e^{t/\tau}\int_0^t(\mathbf{A}-\mathbf{A_0})e^{-t'/\tau}dt'-\int_0^t(\mathbf{A}-\mathbf{A_0})dt'\right]$$

$$=\;\tau\dot{\mathbf{T}}(t)-\frac{q}{mc}\int_0^t(\mathbf{A}-\mathbf{A_0})dt' \qquad \text{(VII-12)}$$

In order to make clear this expression, it is necessary to adopt a precise hypothesis concerning the variations of the vector potential. *We will assume* that the latter can be represented by a succession of Dirac distributions, respectively centred at times t_k. Owing to the isotropy of the universe, the time-average of these variations is equal to zero so that we can choose $\mathbf{A_0}=0$ at $t=0$.

Consequently

$$\mathbf{A}\left(\frac{t}{\tau_0}\right)-\mathbf{A_0}=\mathbf{A}\left(\frac{t}{\tau_0}\right)=\sum\mathbf{A}_k\delta\left(\frac{t-t_k}{\tau_0}\right) \qquad \text{(VII-13)}$$

τ_0 being the time unit.

Given that $\delta(ax) = a^{-1}\delta(x)$, we can write

$$A\left(\frac{t}{\tau_0}\right) = \sum A_k \delta\left(\frac{t - t_k}{\tau}\right).\frac{\tau_0}{\tau} = \frac{\tau_0}{\tau}A\left(\frac{t}{\tau}\right) \tag{VII-14}$$

Then

$$\dot{T}(t) = \frac{q\tau_0}{mc\tau}e^{t/\tau}\int_0^{t/\tau} A\left(\frac{t'}{\tau}\right)e^{-t'/\tau}d\left(\frac{t'}{\tau}\right) \tag{VII-15}$$

The average value of A being equal to zero, from (VII-15) it results that the average values over a sufficiently long time interval $(0,t)$ of $\dot{T}^2$ and T^2 are respectively of the following form

$$\left[\dot{T}^2\right]_0^t = \left(\frac{q\tau_0}{mc\tau}\right)^2 f\left(\frac{t}{\tau}\right) \quad \text{and} \quad \left[T^2\right]_0^t = \left(\frac{q\tau_0}{mc}\right)^2 g\left(\frac{t}{\tau}\right) \tag{VII-16}$$

f and g being functions depending on the average universe field properties but not on the nature of the particle.

Consequently

$$\left[T^2.\overline{(m\dot{T})^2}\right]_0^t = G\left(\frac{t}{\tau}\right) \tag{VII-17}$$

$G(t / \tau)$ being a function depending on the properties of the universe field only.

Given that the oscillations of the field are completely damped before the beginning of the following burst, over the life assigned to the burst under consideration, the function $G (t / \tau)$ has reached a limiting value which depends on the average properties of the universe field only.

In the mean, over a sufficiently great number of bursts, we will put

$$\overline{T^2.(m\dot{T})^2} = 9K^2 \approx (c\tau_0^2)^2.\left(\overline{A^2}\right)^2 \tag{VII-18}$$

K being a constant, independent of the system and of the nature of the particle, and $\overline{A^2}$ the average value of the A_k^2's.

Consequently, for each of the components $u = x, y, z$, we have

$$\overline{T_u^2.(m\dot{T}_u)^2} = K^2 \tag{VII-19}$$

On the other hand, given the assumed stability of the average properties of the universe field, according to (VII-12), over each burst, *in the mean* , the total displacement δT and the variation of the velociy $\delta \dot{T}$ are such as

$$\delta T = \tau \delta \dot{T} \qquad\qquad\qquad \text{(VII-20)}$$

If we consider another system moving with respect to the first one with a certain velocity, the average number of bursts received during the time unit is the same. The number of bursts encountred by the system during its motion , indeed, is increasing while that of bursts which hit it, is decreasing of the same quantity, so that the average value $\overline{A^2}$ is unchanged, thus that the constant K remains the same.

First consequences

Starting from an arbitrary point M_0 of space $\{r\}$, under the effect of the bursts received from the universe field, the particle describes a very complicate trajectory Γ which, under the condition (we assume to be realized) that the system does not disintegrate, passes in the neighborhood of its starting point again. Given the latter is arbitrary, the trajectory Γ *passes in the vicinity of all* (or almost all) *points of space* whatever the initial conditions may be.

On the other hand, given the geometrical structrure of the trajectories, their derivatives are not continuous at every point, which precludes us to define the instantaneous velocity of the particle. Only average values over finite time intervals can be obtained.

This result in a manner justifies the hypothesis of Feynman who was the first to try (partly at least) to come back to a space-time representation of Quantum Mechanics by introducing trajectories which precisely would not possess continuous derivatives. Let us recall as a reminder that this idea has been recently developed by generalizing the relativity principle to a non differentiable space-time which in addition would be fractal in order to find the quantum formalism again, Schrödinger's equation especially [4] .

The loss of memory of the initial conditions which results from the geometrical structure of Γ entails that, after a sufficiently long time, all the possible trajectories Γ are *equivalent* . Consequently, to obtain the average value of a given dynamical property G , it is sufficient to consider *one* arbitrary trajectory Γ , the average value $\overline{G}$ becoming stable from a sufficiently long time. This is typically the character of a *quasi-ergodic* process, as we have concluded in Chapter V.

Another extremely important consequence is the absence of *first integrals* . Indeed if a property G was constant on a trajectory Γ, it would be constant at all the points of space, which would be absurd. Consequently, all the dynamical properties do fluctuate versus time around a well determined average value $\overline{G}$. This is true, in particular, for energy.

Owing to the fact that we can arbitrarily choose the time origin, we can write for the average value over a sufficiently long time

$$\overline{G} = \frac{1}{T}\int_0^T G(t)dt = \frac{1}{T}\int_\theta^{T+\theta} G(t+\theta)dt = \frac{1}{T}\int_\theta^{T+\theta}[G(t)+\theta G'(t)+...]dt$$

i.e. $\quad \overline{G} = \overline{G} + \theta\overline{G'} + \frac{\theta^2}{2}\overline{G''}+...$ \hfill (VII-21)

so that the average value of the derivatives with respect to time of any dynamical property is equal to zero. Likewise, we have

$$\int_0^T G(t)dt = \int_{-T}^0 G(t)dt = -\int_0^T G(-t)d(-t) = \int_0^T G(-t)dt$$

i.e. $\quad \overline{G(-t)} = \overline{G(t)}$ \hfill (VII-22)

In particular, the average value of an odd function of time is equal to zero (e.g. $\dot{\mathbf{r}}$, $\ddot{\mathbf{r}}$, $\mathbf{r}\ddot{\mathbf{r}}$,...) . It follows that the average value of the force $\mathbf{F}$ is equal to that of $m\ddot{\mathbf{r}}$, which is a weaker form of the fundamental principle of classical dynamics, namely

$$\overline{\mathbf{F}} = m\overline{\ddot{\mathbf{r}}}$$ \hfill (VII-23)

instead of $\mathbf{F} = m\ddot{\mathbf{r}}$ which is valid at any time.

Energy balance

Let us multiply (VII-6) by $\dot{r}$ and calculate the corresponding time-average value on an arbitrary trajectory Γ. We obtain

$$-m\tau\overline{\dddot{r}\dot{r}} + m\overline{\ddot{r}\dot{r}} = \overline{F\dot{r}} + q\overline{\dot{r}E} \tag{VII-24}$$

i.e. according to what we have seen in the preceding section for the average values

$$m\tau\overline{\ddot{r}^2} = q\overline{\dot{r}E} \tag{VII-25}$$

The quantity $m\tau\overline{\ddot{r}^2} = -m\tau\overline{\dddot{r}\dot{r}}$ is the average *radiated* power and $q\overline{\dot{r}E}$ the average *absorbed* power. Consequently, in the mean, the radiated power is exactly balanced by the absorbed one. We will notice that relationship (VII-25) remains valid even if we take the magnetic term into account, i.e. if the relativistic corrections are introduced. As expected, this relationship is invariant under a galilean transformation.

Moreover, relation (VII-25) makes the correlation which exists between $\dot{r}$ and E through the damping force appear. This result is general. For instance, for the harmonic oscillator excited by a sinusoidal force

$$-m\tau\dddot{x} + m\ddot{x} + kx = aq\cos\omega t = qE \tag{VII-26}$$

we obtain

$$m\tau\overline{\ddot{x}^2} = q\overline{\dot{x}E} = \frac{1}{2}\frac{a^2q^2m\tau\omega^4}{(m\omega^2 - k)^2 + m^2\tau^2\omega^6} \tag{VII-27}$$

i.e. on the resonance $(k = m\omega^2)$, $a^2q^2/2m\omega^2\tau = 3a^2c^3/4\omega^2$ which is independent of the nature of the particle. We will remark that the energy exchanges are the most important precisely at the resonance.

The virial theorem

Let us multiply (VII-6) by r and calculate the corresponding time-average value. We obtain

$$-m\tau\overline{\dddot{r}r} + m\overline{\ddot{r}r} = \overline{rF} + q\overline{rE} \tag{VII-28}$$

Now $\overline{\dddot{r}r} = 0$ and $\overline{\ddot{r}r} = -\overline{\dot{r}^2}$. On the other hand, $\overline{rE} = -\frac{1}{c}\overline{r\dot{A}} = \frac{1}{c}\overline{\dot{r}A} = 0$

owing to the fact that $\dot{\mathbf{r}}\mathbf{A}$ is an odd function of time. From which

$$m\overline{\dot{r}^2} + \overline{r\mathbf{F}} = 0 \tag{VII-29}$$

which is the virial relationship on the trajectory Γ under consideration. It is to notice that the damping coefficient and the properties of the universe field do not intervene in this relationship.

Effect of a virtual deformation of the trajectory

Let us consider a *virtual* trajectory $\tilde{\Gamma}$ deduced from an actual trajectory Γ by a very small arbitrary modification of the position and of the speed of the particle. More precisely, we will assume that for this trajectory, we have

$$\tilde{\mathbf{r}} = \mathbf{r} + \mathbf{\eta} \tag{VII-30}$$

with $|\mathbf{\eta}| < \varepsilon\sqrt{\overline{\mathbf{r}^2}}$ and $|\dot{\mathbf{\eta}}| < \varepsilon\sqrt{\overline{\dot{\mathbf{r}}^2}}$, ε being an infinitely small of the first order.

The average energy on Γ which is equal to

$$\overline{E} = \frac{1}{2}m\overline{\dot{r}^2} + \overline{U(\mathbf{r})} \tag{VII-31}$$

becomes

$$\overline{\tilde{E}} = \frac{1}{2}m\overline{(\dot{\mathbf{r}}+\dot{\mathbf{\eta}})^2} + \overline{U(\mathbf{r}+\mathbf{\eta})} \tag{VII-32}$$

U being the potential energy. Consequently

$$\overline{\tilde{E}} = \overline{E} + m\overline{\dot{\mathbf{r}}\dot{\mathbf{\eta}}} + \frac{m}{2}\overline{\dot{\mathbf{\eta}}^2} + \overline{\mathbf{\eta}\,\mathrm{grad}U} + \frac{1}{2}\sum_u\sum_v\overline{\frac{\partial^2 U}{\partial u \partial v}\eta_u\eta_v} \tag{VII-33}$$

$(u = x, y, z)$. Given that $\mathbf{r}$ and $\mathbf{\eta}$ are independent from one another $\overline{\mathbf{\eta}\,\mathrm{grad}U} = \overline{\mathbf{\eta}}.\overline{\mathrm{grad}U} = -\overline{\mathbf{\eta}}.\overline{\mathbf{F}} = 0$. Thus

$$\overline{\tilde{E}} - \overline{E} = \frac{m}{2}\overline{\dot{\mathbf{\eta}}^2} + \frac{1}{2}\sum_u\sum_v\overline{\frac{\partial^2 U}{\partial u \partial v}.\eta_u\eta_v} + ... \tag{VII-34}$$

which shows that the variation of the average energy is of the second order with respect to any virtual variation of $\mathbf{r}$ and $\dot{\mathbf{r}}$. In other words, the average energy is *stationary* with respect to any virtual variation of the trajectory.

The Hellmann-Feynman theorem

Let us consider a system whose potential energy is of the following form

$$U = U_0 + \delta U \qquad \text{(VII-35)}$$

δU being a perturbation of the first order.

Let x be the coordinate for the system assumed to be one-dimensional, and x_0, that for the unperturbed system (i.e. corresponding to $\delta U = 0$). We have

$$U(x) = U_0(x) + \delta U(x) = U_0(x_0) + \delta x_0 \frac{\partial U}{\partial x_0} + \delta U(x_0) \qquad \text{(VII-36)}$$

with $x = x_0 + \delta x$.

Likewise, for the kinetic energy

$$T(\dot{x}) = T_0(\dot{x}) = T_0(\dot{x}_0) + \delta \dot{x}_0 \frac{\partial T_0}{\partial \dot{x}_0} \qquad \text{(VII-37)}$$

Thus, for the total energy

$$\overline{E} = \overline{E_0} + \overline{\delta U(x_0)} \qquad \text{(VII-38)}$$

since, in the unperturbed system, according to the previous theorem, we have

$$\delta E_0 = \delta \dot{x}_0 \overline{\frac{\partial T_0}{\partial \dot{x}_0}} + \delta x_0 \overline{\frac{\partial U}{\partial x_0}} = 0 \qquad \text{(VII-39)}$$

Hence

$$\delta \overline{E} = \overline{\delta U} \qquad \text{(VII-40)}$$

which is just the Hellmann-Feynman theorem for a system whose *potential* energy depends on a parameter λ

$$\frac{d\overline{E}}{d\lambda} = \overline{\left(\frac{\partial U}{\partial \lambda} \right)} \qquad \text{(VII-41)}$$

This result can also be stated as follows : The variation of the average energy under the effect of a perturbation V of the potential energy is given, to the first order in V , by the average value of V on the unperturbed trajectory.

The charged harmonic oscillator

The equation

$$-\tau\dddot{x} + \ddot{x} + \omega^2 x = \frac{q}{m}E_x \qquad (\omega^2 = k/m) \tag{VII-42}$$

which governs the motion of such an oscillator can be written as follows, by putting $\theta = t/\tau$

$$-\frac{d^3x}{d\theta^3} + \frac{d^2x}{d\theta^2} + \omega^2\tau^2 x = \frac{q\tau^2}{m}E_x \tag{VII-43}$$

This equation shows that the various properties of such an oscillator differ from the corresponding ones of the free particle by the presence of the term in $\omega\tau$. This dimensionless product can be considered as being a parameter which depends on the strength of the oscillator, and with respect to which the solution of (VII-42) can be discussed.

According to (VII-16) the variations of the position over a sufficiently long time are such as

$$\overline{x^2} = \left(\frac{q}{m}\right)^2 \gamma(\omega\tau) \tag{VII-44}$$

$\gamma(\omega\tau)$ being a function which can be developped in series of the negative and positive powers of $\omega\tau$

$$\gamma(\omega\tau) = ... + a_{-2}(\omega\tau)^{-2} + a_{-1}(\omega\tau)^{-1} + a_0 + a_1(\omega\tau) + ... \tag{VII-45}$$

the coefficients a depending on the average properties of the universe field only.

According to the virial theorem,

$$m\omega^2 \overline{x^2} = \frac{\overline{p_x^2}}{m} = \overline{E} \tag{VII-46}$$

Thus

$$\begin{cases} \overline{x^2} = \frac{\tau}{m}\gamma(\omega\tau) \\ \overline{p_x^2} = m\omega^2\tau.\gamma(\omega\tau) \end{cases} \tag{VII-47}$$

Given that the average values of x and p_x are equal to zero, it results from (VII-47) that

$$\Delta x.\Delta p_x = (\omega\tau).\gamma(\omega\tau) \qquad\qquad (\text{VII-48})$$

Δx and Δp_x being the variances of x and p_x respectively.

In order to determine the function γ we will consider two limiting cases:

i. k tends to zero. The particle is not submitted to the restoring force. The vibration energy $\bar{E} = m\omega^2 \overline{x^2} = \tau^{-1}(\omega\tau)^2 \gamma(\omega\tau)$ tends to zero, which involves that function γ reduces as follows

$$\gamma(\omega\tau) = \frac{a_{-1}}{\omega\tau} + a_0 + a_1(\omega\tau)+... \qquad\qquad (\text{VII-49})$$

As expected, we verify that $\overline{x^2}$ tends to infinity.

ii. k tends to infinity. The particule becomes at rest. Consequently $\overline{x^2}$ tends to zero, so that $\gamma(\omega\tau)$ reduces to the first term of (VII-49), i.e.

$$\gamma(\omega\tau) = \frac{a_{-1}}{\omega\tau} \qquad\qquad (\text{VII-50})$$

Consequently

$$\Delta x.\Delta p_x = a_{-1} \qquad\qquad (\text{VII-51})$$

In order to determine the universal constant a_{-1}, we will consider the limiting case $k \to 0$. Over the time interval $(0,t\,)$ corresponding to n bursts, the displacement x can be decomposed into elementary contributions T_i corresponding to the successive bursts. Let be t_k the time of the beginning of the k-th burst. Owing to the fact that the bursts do not overlap, $T_i(t)$ is equal to zero if $t < t_i$ and $t > t_{i+1}$. Consequently

$$\int_0^t x^2(t)dt = \sum_i \int_{t_i}^{t_{i+1}} T_i^2(t)dt = n\bar{\tau}\overline{T^2} \qquad\qquad (\text{VII-52})$$

$\overline{T^2}$ being the average value of T_i^2 for a sufficiently great number of bursts and $\bar{\tau}$ the average life of a burst.

It follows that, over a sufficiently long time $(t = n\bar{\tau})$, the average value of x^2 is

$$\overline{x^2} = \overline{T^2} \qquad\qquad (\text{VII-53})$$

Likewise for the velocity, so that, according to (VII-19), we obtain

$$\Delta x . \Delta p_x = K \tag{VII-54}$$

In other words, the product of the quadratic dispersions of the position and of the velocity does not depend on the nature of the oscillator but of the average properties of the universe field only.

Concerning the properties themselves of the oscillator under consideration, we obtain

$$\begin{cases} \overline{x^2} = \dfrac{K}{m\omega} \\ \overline{p_x^2} = m\omega K \\ \overline{E} = K\omega \end{cases} \tag{VII-55}$$

All these results can be generalized without difficulty to the case of the tridimensional harmonic oscillator.

Case of complex particles

Up to now we have characterized the particle by its mass and its charge which is assumed to be different from zero. Strictly speaking, our model applies to fundamental charged particles (electrons, muons, quarks), as well as to protons and other nuclei. But it is well known that quantum mechanics applies also to neutral particles (e.g. neutrons) and to vibrational and rotational molecular systems. That arises from the fact that these systems are neutral by compensation only.

In order to understand the phenomenon, let us consider a complex systems built up on particles i (e.g. the quarks for a neutron) of mass m_i . The set of these particles i making a rigid body, for each of them we can write

$$-\tau_i \ddot{\mathbf{r}} + \ddot{\mathbf{r}} = \frac{\mathbf{F(r)}}{m_i} + \frac{q_i}{m_i} \mathbf{E}(t) \tag{VII-56}$$

(the coordinates $\mathbf{r}_i$ of the particles i have been replaced by their common value $\mathbf{r}$. This reduction of the problem dispenses us from explicitely introducing the internal bond forces between the particles) . From which, in carrying out the averaging of all the particles

$$-\bar{\tau}\dddot{\mathbf{r}} + \ddot{\mathbf{r}} = \frac{\overline{\mathbf{F(r)}}}{m'} + \overline{\left(\frac{q}{m}\right)}\mathbf{E}(t) \qquad \text{(VII-57)}$$

with $\dfrac{1}{m'} = \overline{\left(\dfrac{1}{m}\right)}$. This equation is formally identical with that which would

correspond to a particle of mass m ' and of charge q ' such as

$$q'^2 = \frac{\overline{q^2 / m}}{\overline{1 / m}} = \frac{\sum q_i^2 / m_i}{\sum 1 / m_i} \qquad \text{(VII-58)}$$

Thus, this charge is never equal to zero. Consequently, our model applies with the same constant K , for any particle and for any molecular system of finite size.

Neutrino which is neutral by nature and not by compensation, does not interact with the universe field.

The angular momentum

To the motion on Γ, it corresponds a kinetic momentum $\mathbf{M}$. Given that this momentum $(\mathbf{r} \wedge m\dot{\mathbf{r}})$ is an odd function of time, according to what

we have seen, its time-average projection along an arbitrary direction z is equal to zero, whatever the system under consideration may be.

The case of hydrogen atom is typical with this regard. In the absence of the universe field, according to the kinetic momentum theorem $(\mathbf{r} \wedge \mathbf{F} = d\mathbf{M} / dt)$, the momentum would be constant (in modulus and in direction). Under the effect of the universe field, the modulus fluctuates and the vector successively orientates towards all the directions of space in a random manner so that $\overline{M_z} = 0$.

Effect of a magnetic field

Let us assume that the system is immersed within a constant magnetic field H_0. In the equation which governs the motion of a particle, the supplementary term $\dfrac{q}{m}\dfrac{\dot{r}}{c}\wedge H_0$ appears. Now this term has not to be taken into account to determine the trajectory Γ owing to the fact this trajectory corresponds to a non-relativistic level. Consequently, the trajectory Γ is unchanged. We will come back subsequently (Chap. IX) to the problem when we will have introduced the relativistic corrections.

Conclusion

It emerges from this discussion that our model, based on the balance between the energy radiated by the system and that received from the rest of the universe, leads to general results quite consistent with the qualitative conclusions we have obtained in the previous chapters. More still, for the harmonic oscillator, we obtain an energy and a product $\Delta x.\Delta p_x$ which can be identified with the quantal results by putting $K = \hbar / 2$. Nevertheless, we will notice that, if effectively we find the Heisenberg relationship again, the meaning of the latter is fundamentaly different from that claimed by the orthodox quantum mechanics. Indeed this relationship does not involve any impossibility connected with the nature of matter, to know the position and the speed of the particles simultaneously with an arbitrarily great accuracy.

A point remains to examine, namely the ability for our model to find the quantum formalism again. The problem will be examined in the following chapter.

References

[1] This chapter takes again the ideas and the results previously published in : A. Julg, *Ann. Fond. Louis de Broglie* **16** (1991) 321 and A. Julg in *Waves and Particles*

in Light and Matter, A. van der Merwe and A. Garuccio eds (Plenum, New York) 1994, p.337 , with some modifications and supplementary details.

[2] L. Landau, F. Lifchitz, *Physique Théorique II (Théorie des champs)* , (Mir, Moscou) 3ème éd. 1970, p. 280.

[3] J.D. Jackson, *Classical Electrodynamics* (Wiley, New York) 2th ed., 1975, p.560.

[4] L. Nottale, in *Quantum Mechanics, diffusion and chaotic fractals* , M.El Nashie, O. Rössler, I. Prigogine Eds. (Pergamon Press, New York) 1995 and references therein.

VIII

Connection with the quantum formalism

Transcription into an operator formalism

The model we have constructed in the previous chapter, is essentially based on the classical methods of the differential calculus, while quantum mechanics utilizes an operator formalism. If the two approaches allow to describe the physical phenomena with the same efficiency, certainly it does exist a connection between the two manner of treating the problem.

In order to join these two approaches, we will start from the principle [1] according to which the result of a measurement can be represented by the usual integral

$$\overline{G} = \int \psi^* \hat{G} \psi \, dv = \langle \psi \hat{G} \psi \rangle \qquad \text{(VIII-1)}$$

where $\hat{G}$ is a linear operator associated with the property G , and ψ a function of space such as the square of its modulus is equal to the density $\rho(\mathbf{r})$ in the space $\{\mathbf{r}\}$

$$\psi^* \psi = \rho(\mathbf{r}) \qquad \text{(VIII-2)}$$

which involves that the function ψ is normalized to unit, i.e. that we have $\int \psi^* \psi \, dv = 1$. This function tends necessarily to zero when $\mathbf{r}$ tends to the infinity. $\overline{G}$ (VIII-1) being a real number, it is sufficient that the operator $\hat{G}$ be *hermitian* .

For a property depending on $\mathbf{r}$ only, we can put

$$\hat{G} = G \qquad \text{(VIII-3)}$$

Indeed,

$$\overline{G} = \int G(\mathbf{r}) \rho(\mathbf{r}) \, dv \qquad \text{(VIII-4)}$$

More generally, for a property $G(x, p_x, ...)$ we will put

$$\hat{G}(x, p_x, ...) = G(x, \hat{p}_x, ...) \tag{VIII-5}$$

$\hat{p}_x, \hat{p}_y, \hat{p}_z$ being the operators associated respectively with the components p_x, p_y, p_z of the momentum of the particle. As the properties p_x, p_y, p_z themselves, the operators $\hat{p}_x, \hat{p}_y, \hat{p}_z$ must be invariant under the translation $x \rightarrow x + a, ...$, which involves that the operators $\hat{p}_u$ ($u = x$, y, z) are of the following form

$$\hat{p}_u = \sum_k C_k \frac{\partial^k}{\partial u^k} \tag{VIII-6}$$

k being an integer different from zero.

According to (VIII-5), the average kinetic energy for an one-dimensional system is

$$\overline{T} = \frac{1}{2m} \int \psi^*(x) (\hat{p}_x)^2 \psi(x) dx \tag{VIII-7}$$

Let us make an infinitely small translation a along x . This energy becomes

$$\overline{T_a} = \int \psi^*(x + a)(\hat{p}_x)^2 \psi(x + a) dx \tag{VIII-8}$$

Now

$$\psi(x + a) = \psi(x) + a \frac{\partial \psi}{\partial x}(x) \tag{VIII-9}$$

As expected, we verify that $\overline{T_a} = \overline{T}$.

On the other hand, p_u being an odd function of u in classical mechanics, k must be an odd integer. Practically, we will limit the development to the first order.

Consequently, for the kinetic energy of the particle, we obtain

$$\overline{T} = \frac{C_1^2}{2m} \int \psi^* \frac{\partial^2 \psi}{\partial x^2} dx = -\frac{C_1^2}{2m} \int \left| \frac{\partial \psi}{\partial x} \right|^2 dx \tag{VIII-10}$$

The kinetic energy being positive, it is necessary that C_1 is an imaginary number, which leads us to the operator associated with p_x

$$\hat{p}_x = \frac{C}{i} \frac{\partial}{\partial x} \tag{VIII-11}$$

and the analogous ones for the other components, with the same constant C because the space is isotropic.

The Schrödinger equation

The average energy is

$$\overline{E} = \left\langle \psi \left| -\frac{C^2}{2m} \nabla^2 + U \right| \psi \right\rangle = \langle \psi \hat{H} \psi \rangle \qquad \text{(VIII-12)}$$

U being the potential energy, and $\hat{H}$ the operator associated with the energy, i.e. the *hamiltonian* operator.

Now we have seen that the average energy $\overline{E}$ is stationary with respect to any virtual variation of the trajectory Γ described by the particle, i.e. to the density $\rho(\mathbf{r})$ (VIII-2) . This energy, therefore, is stationary with respect to any variation of ψ itself

$$\delta \overline{E} = \langle \delta \psi \hat{H} \psi \rangle + \langle \psi \hat{H} \delta \psi \rangle = 0 \qquad \text{(VIII-13)}$$

which involves that ψ is an eigenfunction of $\hat{H}$, i.e. that we have

$$\hat{H}\psi = \overline{E}\psi \qquad \text{(VIII-14)}$$

The proof of this result is the following. Let us develop ψ and its variation $\delta\psi$ as linear combinations of the eigenfunctions ϕ_k of the operator $\hat{H}$, corresponding to the eigenvalues E_k . We obtain

$$\begin{cases} \psi = \sum_k c_k \phi_k \\ \delta\psi = \sum_j \alpha_j \phi_j \end{cases} \qquad \text{(VIII-15)}$$

with $\langle \delta\psi . \psi \rangle = \sum_k \sum_j c_k \alpha_j \langle \phi_k \phi_j \rangle = \sum_k c_k \alpha_k = 0$ (VIII-16)

owing to the fact that the varied function $\psi + \delta\psi$ must remain normalized.

Hence

$$\left\langle \sum_j \alpha_j \phi_j \left| \hat{H} \right| \sum_k c_k \phi_k \right\rangle = \sum_j \sum_k \alpha_j c_k E_k \langle \phi_j \phi_k \rangle = \sum_k c_k \alpha_k E_k \quad \text{(VIII-17)}$$

which is equal to zero by virtue of (VIII-14). Taking (VIII-16) into account, we obtain

$$\sum_k c_k \alpha_k (E_k - \overline{E}) = 0 \qquad \text{(VIII-18)}$$

Given that α_k is arbitrary, it is necessary that

$$c_k (E_k - \overline{E}) = 0 \qquad (\forall k) \qquad \text{(VIII-19)}$$

which requires that ψ is identical with the eigenfunction ϕ_k corresponding to the eigenvalue $E_k = \overline{E}$.

Thus, we formally find the Schrödinger equation again, but, here, $\overline{E}$ is the *average value* of the energy which, in fact, is the sole value able to be reached by experiment. We will notice that we do not obtain a unique state, but a set of states for which energy is stationary. Presently, we will only take interest to the state whose energy is the lowest, the *ground state*. The other states, called *excited states*, will be studied subsequently (Chap. XI).

In order to determine the constant C, we can, for instance, integrate the equation (VIII-14) for an harmonic oscillator. For the ground state, we obtain $\overline{E} = \dfrac{C}{2} \omega$, which, by identifying with the result obtained in the previous chapter, gives

$$C = 2K \qquad \text{(VIII-20)}$$

Consequently, to find the quantum formalism again, we must put

$$2K = \hbar \qquad \text{(VIII-21)}$$

Coming back to the harmonic oscillator

Let us consider the harmonic oscillator governed by the following equation

$$m\ddot{x} + m\omega^2 x = 0 \qquad \text{(VIII-22)}$$

The first integral of which is

$$\frac{1}{2m} p^2 + \frac{1}{2} m\omega^2 x^2 = E \qquad \text{(VIII-23)}$$

Within the representation (x , p) , i.e. within the phase space, a well determined trajectory γ corresponds to each of the values of E . These various trajectories γ are not secant. Along each of them, the values of x and p are determined by the energy. The general solution, indeed, is of the form

$$x = a\sin(\omega t - \varphi) \ , \ \ p = ma\omega\cos(\omega t - \varphi) \tag{VIII-24}$$

with $E = \dfrac{1}{2}ma^2\omega^2$, hence

$$x = \sqrt{\frac{2E}{m\omega^2}}\sin(\omega t - \varphi) \ \text{ and } \ p = \sqrt{2mE}\cos(\omega t - \varphi) \tag{VIII-25}$$

Let us assume that it exists a certain probability law which would allow to obtain the trajectory γ, i.e. a certain energy E : P (E). As it is easy to see, along this trajectory γ of energy E , we have

$$\overline{\left(x^2\right)}_\gamma = \frac{E}{m\omega^2} \ , \ \ \overline{\left(x^4\right)}_\gamma = \frac{3E^2}{2(m\omega^2)^2} \ , \ \ \overline{\left(p^2\right)}_\gamma = mE \ , \ ... \tag{VIII-26}$$

Thus, for the whole of the trajectories γ

$$\begin{cases} \overline{x^2} = \dfrac{1}{m\omega^2}\int EP(E)dE = \dfrac{\overline{E}}{m\omega^2} \\[2mm] \overline{x^4} = \dfrac{3\overline{E^2}}{2(m\omega^2)^2} \\[2mm] \overline{p^2} = m\overline{E} \end{cases} \tag{VIII-27}$$

with $\overline{E} = \int EP(E)dE$.

Given that $\overline{E} = \hbar\omega/2$, for $\overline{x^2}$ and $\overline{p^2}$ we obtain the quantal result again, namely

$$\overline{x^2} = \frac{\hbar}{2m\omega} \ , \ \ \overline{p^2} = \frac{\hbar\omega m}{2} \ , \ \text{i.e. } \ \overline{x^2}.\overline{p^2} = \frac{\hbar^2}{4} \tag{VIII-28}$$

But, for $\overline{x^4}$, where the average value of E^2 (VIII-27) appears, the problem is different. In the orthodox quantum mechanics, E is a constant, so that $\overline{E^2} = (\overline{E})^2$. Then, for the ground state, $\overline{x^4} = \dfrac{3}{8}\left(\dfrac{\hbar}{m\omega^2}\right)^2$. Now the actual value is two times greater. On the contrary, if we adopt the complete symmetrization of the operators, according to (III-13) , we obtain

$$\overline{E^2} = \left(\overline{E}\right)^2 + \frac{\hbar^2 \omega^2}{4} \qquad \text{(VIII-29)}$$

i.e. for the nth level

$$\left(\overline{E^2}\right)_n = \left(n + \frac{1}{2}\right)^2 \hbar^2 \omega^2 + \frac{\hbar^2 \omega^2}{4} = \left(n^2 + n + \frac{1}{2}\right)\hbar^2 \omega^2 \qquad \text{(VIII-30)}$$

Hence

$$\overline{x^4} = \frac{3}{4}\left(2n^2 + 2n + 1\right)\frac{\hbar^2}{m^2 \omega^2} \qquad \text{(VIII-31)}$$

which is the value deduced from the Schrödinger equation.

This example shows :

i. that the operators have to be symmetrized,

ii. that the quantum description can be obtained by a continuous
distribution of harmonic oscillators.

We will notice that the law $P (E)$ is not made precise. This law is, certainly, not the same for the various states n .

Let us try to find whether such an essentially classical description is general. In this view, we will consider the case of a particle submitted to the following 1D -potential : $U(x) = \frac{1}{2}m\omega^2 x^2 + \sum_{n \neq 2} a_n x^n$.

A calculation, analogous to that we have performed for the harmonic oscillator, is possible only if x is of the following form

$$x = f(E)g(t) \qquad \text{(VIII-32)}$$

In this case, the following generalization of (VIII-23)

$$\frac{m^2 f^2 \dot{g}^2}{2m} + U(fg) = E \qquad \text{(VIII-33)}$$

has to be verified for each trajectories γ , and, consequently, whatever t may be, the energy remaining constant. Then

$$mf^2 \ddot{g}\dot{g} + \frac{\partial U}{\partial g}(fg)\dot{g} = 0$$

i.e. $mf^2(\ddot{g} + \omega^2 g) + \sum_{n \neq 2} n a_n f^n g^{n-1} \equiv 0$

$$(\text{VIII-34})$$

whatever the energy may be, thus whatever f. This relationship can be considered as a polynomial in f, identical to zero. Consequently, its coefficients are equal to zero, i.e.

$$\ddot{g} + \omega^2 g = 0 \quad \text{and} \quad a_{n \neq 2} = 0 \qquad (\text{VIII-35})$$

In other words, the harmonic oscillator is the sole system for which such a calculation procedure, performed in two steps, is possible. This result has to be compared with the fact that in quantum mechanics the phase space (q , p) is convenient for the harmonic oscillator only. It is, indeed, well-known that , excepted for the harmonic oscillator, the density in this space is not ever strictly positive. It can, even, be negative. A space of greater dimension is, nevertheless, not excluded for interpreting the phenomenon [2] .

The rigid rotator in a plane

The rigid rotator is, of course, an idealization which does not occur in nature. Its case, nevertheless, is worth examining in order to compare our model with the conventional Quantum Mechanics.

Let I be its inertia momentum with respect to its rotation axis z . Classically, the z -component of the angular momentum M_z is equal to $I\omega$ and the rotation energy to $\frac{1}{2}I\omega^2$, ω being the rotation angular frequency, so that the classical energy is

$$E_{class.} = \frac{1}{2I}M_z^2 \qquad (\text{VIII-36})$$

from which the conventional quantum result

$$E = \frac{1}{2I} \langle M_z^2 \rangle \tag{VIII-37}$$

In our model Eq(VIII-36) becomes

$$\overline{E} = \frac{1}{2I} \overline{M_z^2} \tag{VIII-38}$$

According to what we have seen in Chapter VI for a free particle, the average value of the square of the angular momentum arising from the universe field is the following

$$\overline{M^2} = \overline{(xp_y - yp_x)^2} + \overline{(yp_z - zp_y)^2} + \overline{(zp_x - xp_z)^2}$$
$$= (\overline{x^2} . \overline{p_y^2} - 2\overline{xp_x} . \overline{yp_y} + \overline{y^2} . \overline{p_x^2}) + ... \tag{VIII-39}$$
$$= 6\overline{x^2} . \overline{p_y^2} = 6\overline{x^2} . \overline{p_x^2} = 6\frac{\hbar^2}{4} = \frac{3}{2}\hbar^2$$

In the rigid rotator, the particle is constrained to move on a circle of z- axis so that only the z -component of the momentum is involved. Consequently, given the isotropy of space, we have

$$\overline{M_z^2} = \frac{\hbar^2}{2} \tag{VIII-40}$$

so that the rotation energy is

$$\overline{E} = \frac{\hbar^2}{4I} \tag{VIII-41}$$

Hence in our model, contrary to what occurs in the conventional Quantum Mechanics, the rigid rotator in a plane, in its ground state, is never at rest. Just as the harmonic oscillator, it is unceasingly moving under the effect of the universe field, rotating alternately in a direction and in the other.

Let us recall that Stochastic Electrodynamics leads to the same result through a completely different way [3] .

But the most important result which emerges from this study is that the value obtained for the energy (VIII-41) quantitatively agrees with the one which would be deduced from a quantum formalism in which the operators would be completely symmetrized (III-26) .

The time-dependent Schrödinger equation

Let us consider a system whose potential energy depends on the time sufficiently slowly for the ergodicity to be considered as being reached at any time, i.e. for $\overline{E} = \langle \Psi \hat{H} \Psi \rangle$ to be verified at any time. The wavefunction depends on t , the latter playing the role of a slowly variable parameter. Given the normalization condition, we can write

$$\Psi(t + \delta t) = \exp(-i\hat{B}\delta t)\Psi(t) = \Psi - i\hat{B}\Psi\delta t \qquad \text{(VIII-42)}$$

the operator $\hat{B}$ being real and hermitian.

From which it results

$$\int (\Psi^* + i\hat{B}\Psi^*\delta t)|\hat{H} + \delta\hat{H}|(\Psi - i\hat{B}\Psi\delta t)dv = \overline{E} + \delta\overline{E} \qquad \text{(VIII-43)}$$

i.e. according to the Hellman-Feynman theorem (VII-14)

$$\int \Psi^* \hat{H}\hat{B}\Psi dv = \int \Psi \hat{H}\hat{B}\Psi^* dv \qquad \text{(VIII-44)}$$

which involves that the product $\hat{H}\hat{B}$ is hermitian, thus that $\hat{H}$ and $\hat{B}$ commute. Given that p and r are independent variables, on the analogy with the structure of the hamiltonian operator, we will look for a solution in the form

$$\hat{B} = \hat{F}(\hat{p}) + \hat{G}(r,t) \qquad \text{(VIII-45)}$$

Now $\left[\hat{H}, \hat{B}\right] = 0$ involves $\left[\hat{p}^2, \hat{G}\right] = \left[\hat{F}, V\right]$. Consequently, if V is multiplied by λ , $\hat{G}$ is also multiplied by this factor, so that $\hat{G}$ is proportional to V . Thus

$$\left[\hat{p}^2, V\right] = \left[\hat{F}, V\right] \text{ i.e. } \left[\hat{p}^2 - \hat{F}, V\right] = 0 \qquad \text{(VII-46)}$$

whatever V may be, which involves that $\hat{F} = \hat{p}^2$, from which it follows that $\hat{B}$ is proportional to $\hat{H}$ (This result is not surprising because the hamiltonian operator potentially contains all the informations concerning the system).

Thus, by homogeneity, we obtain

$$i\hbar \frac{\partial \Psi}{\partial t} = A\hat{H}\Psi \qquad\qquad (VIII-46)$$

A being a dimensionless number. If $A = 1$, we find the time-dependent Schrödinger equation again.

We will notice that the proof of this equation restricts its applicability domain to phenomena whose characteristic times are much greater than the ergodicity times of the systems under consideration. Consequently, this equation is unsuited to describe the phenomena which occur during the electronic transitions which involve times $(10^{-23} - 10^{-25} s)$ very small with respect to the atomic and molecular ergodicity times (See Chap. V) .

Origin of the universality of the Schrödinger equation

At the present time, everyone agrees that Quantum Mechanics is allowed to be applied to all the particles and to any system built up upon the latter. Such a universality was a priori not obvious. Quantum Mechanics indeed introduces a unique parameter, namely the Planck constant which is essentially connected with the electromagnetic interactions through the fine structure constant $\alpha = e^2 / \hbar c$. Now Quantum Mechanics has been precisely constructed to account for the properties of systems governed by electromagnetic interactions, atoms in particular, so that one could think that for systems in which the weak and strong interactions intervene, as in mesons and hadrons constituted by quarks, supplementary parameters corresponding to these interactions should be introduced. In fact, nothing of the kind. The quantum formalism elaborated at the beginning of the century occurs valid for systems where weak and strong forces intervene [4] , which does not contribute to make the very physical origin of Quantum Mechanics clear.

Our model allows to understand this suitability of Quantum Mechanics for all systems. In our interpretation indeed the Planck constant arises from the electromagnetic interactions between the system under consideration and the rest of the universe. Now the weak and strong interaction forces have very short ranges $(< 10^{-13} cm)$ so that they do not play a role in the interactions with the surrounding medium. Thus no supplementary parameter corresponding to these forces has to be introduced in the equations. Concerning the gravitation forces, the radiated power is so weak that, even for astronomical bodies, their effects upon the motion are completely negligible [5], which dispenses us from introducing a damping term and, consequently, a parameter which would be the equivalent of the Planck constant.

On the other hand, in the proof of the Schrödinger equation we have given, no hypothesis has been made concerning the origin of the potential energy U (VIII-12). All the kinds of interaction can be introduced without the formalism being modified.

At last, let us recall that we have shown (Chapter VII) that all the systems whose the zero charge arises from the balance between the negative and the positive charges of their components, are interacting with the universe field so that such systems are also relevant to Quantum Mechanics.

Thus, although the quantum formalism introduces a unique parameter connected with the electromagnetic interactions, it presents a universal character. The unique limitation of this universality which could exist, would be the case of neutrinos which are fundamental particles neutral by nature. To our knowledge, this point has never pointed out. An experimental confirmation would be a supplementary argument for our model.

Meaning of the quantum formalism

All things considered, it results from the previous calculations that the quantum formalism is a possible algebra (other formalisms can, *a priori* , exist) designed for describing the behavior of a system whose only average values of the various properties are determined, without expliciting the instantaneous behavior of the system clearly. This explains the formal analogy with a statistical set which may be interpreted as the proof of the statistical character of quantum mechanics. In fact, the quantum formalism is only the transcription of this kind of behavior for a single system.

Whatever that may be, the quantum formalism and the reality are two completely different things which have to be carefully distinguished from one another.

Stability of atoms and molecules

The starting point of our interpretation of Quantum Mechanics has been the problem of the stability of hydrogen atom. We have put that the latter arises from the balance between the energy radiated by the electron in its motion around the nucleus and the energy received from the rest of the universe. After which, we have developed a mathematical model implicitely based on the hypothesis that such an equilibrium state is effectively reached and that the system does not disintegrate (i.e. in the case of hydrogen atom that the electron remains linked with the proton) , which has allowed us to find the quantum formalism again. Now is the time to examine the problem a little more closely. The question, indeed, is all the more serious as SED which, as our model, introduces an external field (See Chap I) , encounters a disturbing difficulty with hydrogen atom which is found to be unstable with respect to its components, namely the proton and the electron [6] .

In our model the situation is the following. Under the effect of the impulses arising from the universe field, the electron can jump from an arc of a bonding elliptical trajectory C (See Chap. VII) of negative energy on an anti-bonding hyperbolic trajectory of positive energy, so that, although the time-average value of the universe field is equal to zero, thanks to a conjugation of casual circumstances, the accumulating of such jumps risks to move off the electron from the nucleus in an irreversible manner owing to the fact that the proton-electron attraction tends to zero as r^{-2} (In the harmonic oscillator the restoring force increases as r so that the system remains closed) . Actually, the probability that such an event occurs is extremely weak, but it is not rigorously equal to zero, so that after a sufficiently long time spontaneously the atom will be auto-ionizing. Now that is not what Quantum Mechanics foresees. Moreover, no experiment has observed such a phenomenon.

In fact, the difficulty is only apparent. Indeed, we have seen that our interpretation based on a well-defined physical model, allows to find the Schrödinger equation again precisely by assuming that the system does not disintegrate. Owing to the fact that our model foresees such a disintegration, we must conclude that the Schrödinger equation (VIII-14) is valid only so long as the system remains closed. Consequently, the stability we observe for hydrogen atom and, more generally, for any atom and for any molecule, would be relative given that it is limited in the time. Sooner or later all these systems become auto-ionizing. The inconsistence with experiment is removed if we admit that, at our scale, the lifetimes of these systems are sufficiently great for us to be entitled to consider them as being stable.

The situation would be analogous with the one encountered for the proton itself : This nucleus is considered in Chemistry as well as in Physics as a fundamental particle although the theory of the Great Unification [7] attribuates a finite life time (c.a. 10^{31} years) to it. Still here, all is matter of time!

References

[1] A. Julg, *Annales Fondation Louis de Broglie* , **16** (1991) 321.

[2] A. Julg, P. Julg, *Int. J. Quantum Chem.* , **XXIII** (1978) 483.

[3] T.H. Boyer, *Physical Review D* , **1** (1970) 2257.

[4] W. Buchmüller, S.H.H. Tye, *Phys. Rev. Letters* **45** (1980) 103 , *Phys. Rev.* D, **24** (1981) 132 ; A. Martin, *Phys. Lett.* **100** B (1981) 511.

[5] L. Landau, E. Lifchitz, *Physique Théorique, Théorie des champs* (Mir, Moscou) 1970, p. 431.

[6] L. Pasquera, P. Claverie, *J. Math. Phys.* **23** (1982) 1315.

[7] P. Langacker, *Physics Repport* **72** (1981).

IX

The electron spin

Ambiguity of the notion

If a notion is considered as having specifically a quantum nature, i.e. without any classical equivalent, that is undoubtedly that of spin of particles. In fact, the question is far from being so simple and it is interesting to recall the story of the latter.

One says often that the electron spin has been discovered by Stern and Gerlach in 1921 [1] . Initially, these authors proposed to verify that, according to the quantum theory of this time (at present, we would say the "old quantum theory") , the silver atom exhibits a kinetic momentum arising from the motion of the optical electron, and that this momentum is equal to $h/2\pi$. Effectively, their experiments showed up a magnetic momentum corresponding exactly to this kinetic momentum. But, subsequently, after the Schrödinger works, one realized that the orbital momentum of the electron in the silver atom is equal to zero. From which Uhlenbeck and Goudsmit concluded that the electron exhibits an intrisic kinetic momentum, and consequently a magnetic momentum, independent of its orbital motion [2] . These authors called this momentum "spin momentum" , being more or less conscious that electron is similar to a small sphere spinning about an axis through it.

In 1931, Pauli applied the general quantum theory of the kinetic momenta to spin, which induced him to introduce quantum spin-operators, in a purely formal way, without reference to any classical property contrary to which occurs for the other operators (cf. Chap. III) . From which he deduced that the value of the component s_z of the spin momentum along an arbitrary direction z , is equal to $\pm \hbar / 2$, which required the introduction of the corrective factor g =2, called *gyromagnetic factor* , in order to find the experimental value of the electron magnetic momentum again. Subsequently, Dirac showed that the relativistic quantum mechanics can justify the value of this factor, and improved its value. Owing to the fact that the spin notion appeared without reference to any classical property, and that it did not give the correct value of the magnetic momentum within the framework of a classical theory (which would demand that g is equal to 1) , the opinion was spreading that, as the tunnelling effect, the spin is essentially quantal in nature.

These difficulties, nevertheless, did not discourage the theorists from semi-classical researches. The most interesting approach is that which introduces a quick trembling, called *Zitterbewegung* [3] , around the average trajectory of the particle, which would be responsible for its supplementary intrinsic momentum. Various models have been proposed, in particular, a right- or left-handed helicoidal motion. But, in spite of interesting results, the model remains unfruitful owing to the fact that it does not make the origin of this superimposed motion precise.

For its part, stochastic electrodynamics seems to manage to find the intrisic kinetic momentum again from the fluctuations of the orbital momentum arising from the vacuum field [4] . Nevertheless, it is necessary to introduce the corrective factor 2 in order to find the experimental value of the magnetic momentum again.

In fact, there are two notions which have not to be intermingled, that of spin-momentum, issued from the quantum formalism, and that of the intrinsic momentum of the particle which is physical in nature. Ignoring this duality of interpretation is the cause of all the difficulties. We are going to see how our model allows to understand the problem [5].

The intrinsic kinetic momentum of the electron

The results we have up to now obtained (Chap. VII and VIII) have been deduced from the hypothesis that the velocity of the particle is negligible with respect to that of light ($\beta = v/c \approx 0$). Let us examine what occurs if we take the terms of superior order in β into account.

We have seen (VII-5) that, if we neglect the β^2-terms, the equation of the motion is the following

$$-\tau \ddot{\mathbf{R}} + \ddot{\mathbf{R}} = \frac{\mathbf{F(R)}}{m} + \frac{q}{m}\mathbf{E_R}(t) + \frac{q}{m}\frac{\dot{\mathbf{R}}}{c} \wedge \mathbf{H}(t) \qquad \text{(IX-1)}$$

where $\mathbf{E_R}(t)$ is the value of the electric field at the time t, and at the point $\mathbf{R}$. Within this approximation, the propagation effect has not to be introduced for the magnetic field. This equation generalizes Eq (VIII-6) previously obtained

$$-\tau \ddot{\mathbf{r}} + \ddot{\mathbf{r}} = \frac{\mathbf{F(r)}}{m} + \frac{q}{m}\mathbf{E_r}(t) \qquad \text{(IX-2)}$$

which defines the trajectory Γ.

Let us put

$$\mathbf{R} = \mathbf{r} + \rho \qquad \text{(IX-3)}$$

ρ being the correction arising from the relativistic effects ($\beta \neq 0$) which has to be added to the solution $\mathbf{r}$ (IX-2). By subtracting Eq (IX-2) from Eq (IX-1), we obtain

$$-\tau \ddot{\rho} + \ddot{\rho} = \frac{q}{m}\frac{\dot{\rho}}{c} \wedge \mathbf{H}(t) + \frac{\mathbf{F(r+\rho)} - \mathbf{F(r)}}{m} + \frac{q}{m}\left[\mathbf{E_{r+\rho}}(t) - \mathbf{E_r}(t)\right] \qquad \text{(IX-4)}$$

Thus, by taking ρ as an infinitely small quantity of the first order and by neglecting the contribution brought by the deterministic force with respect to that arising from the universe field

$$-\tau \ddot{\rho} + \ddot{\rho} = \frac{q}{m}\frac{\dot{\rho}}{c} \wedge \mathbf{H}(t) + \frac{q}{m}\frac{\rho_n}{c}\dot{\mathbf{E}}(t) \qquad \text{(IX-5)}$$

This equation describes the motion of the particle within a reference-frame carried on the trajectory Γ with the velocity $|\dot{r}| \ll c$. Let Z be the trajectory within this frame. Owing to the isotropy of the universe field, $\bar{\rho} = 0$ and $\dot{\bar{\rho}} = 0$.

In order to study the motion on Z, we will proceed in the same manner as for the trajectory Γ, i.e. by separating the electric term from the magnetic one

$$\rho = \rho_H + \rho_E \qquad (IX-6)$$

Concerning the magnetic term, the equation

$$-\tau \ddot{\rho}_H + \dot{\rho}_H = \frac{q}{m} \frac{\dot{\rho}_H}{c} \wedge H \qquad (IX-7)$$

shows that the effect of a burst of the universe field is to modify the *direction* of the velocity $\dot{\rho}_H$, but not its modulus, so that, owing to the τ-term, the particle is accelerated up to a limiting velocity, close to that of light. Strictly speaking, Eq (IX-7) is not sufficient. Nevertheless, given that the modulus of the velocity remains constant, the variation of the velocity-vector, perpendicular to the initial velocity and to the field, is weak with respect to the light velocity, so that we can still use Eq (IX-7) to calculate the variation of the velocity.

Given that the order of magnitude of the modulus of $\dot{\rho}_H$ is approximately equal to c, and that the moduli of H and E are equal, the formal analogy with Eq (VII-9) allows to conclude that the variation of the velocity is of about $\sqrt{2K/3}$. The factor $\sqrt{2/3}$ arises from the fact that the speed and the field are randomly oriented the one with respect to the other.

Concerning the electric field, the equation of the motion can be written in an analogous manner

$$-\tau \ddot{\rho}_E + \dot{\rho}_E = \frac{q}{m} \frac{\rho_n}{c} \dot{E} \approx -\frac{q}{m} \frac{\dot{\rho}_n}{c} E \approx -\frac{q}{m} \sqrt{\frac{2}{3}} E \qquad (IX-9)$$

Thus, the variation of the position is equal to $\sqrt{\frac{2}{3} K}$ along E.

To the motion on Z, there corresponds the following kinetic momentum

$$\sigma = \rho \wedge m\dot{\rho} \qquad \text{(IX-10)}$$

The formal analogy between (IX-7) and (IX-9) on the one hand, and (IX-2) on the other, leads for ρ to a relationship analogous to (VII-19). As $\overline{T}$ and $\overline{\dot{T}}$, $\overline{\rho}$ and $\overline{\dot{\rho}}$ are equal to zero. Nevertheless, in the present case, we must take the fact that σ is a vector product into account. The average value of the square of the sine of the directions $\delta\rho$ and $\delta\pi = m\delta\dot{\rho}$ is equal to 3/4 (If we assume that $\mathbf{E}$ is oriented along the x -axis, and that $\mathbf{H}$ is located on the plane $x = 0$, and $\delta\pi$ on a plane perpendicular to $\mathbf{H}$, then $\cos^2(\delta\rho, \delta\pi) = \cos^2(\mathbf{x}, \delta\rho).\sin^2(\mathbf{z}, \mathbf{H})$ whose average value is equal to 1/4). Consequently

$$\overline{\sigma^2} = \frac{3}{4} \times \frac{2}{3} \times \frac{2}{3} \times 9K^2 = 3K^2 = \frac{3}{4}\hbar^2 \qquad \text{(IX-11)}$$

We find the quantum result again, but as an average value. Our calculation makes the deep connection appear between the spin and the Heisenberg relationship. Let us recall that Hestenes [6] which had an inkling of this connection within a framework different from ours, was very sharply critized for that in the past.

The momentum σ has been defined within a reference-frame moving with the speed $\dot{r}$ very inferior to that of light, with respect to a frame bound to the trajectory Γ.

Within this frame we will call the absolute frame, the intrinsic kinetic momentum is $\sigma_{abs} = (r + \rho) \wedge (m\dot{r} + \pi) = (\rho \wedge \pi) +$ terms whose average value is equal to zero. Consequently, everything occurs for the observer, as if, in its actual motion, the particle exhibited a kinetic momentum equal to that defined within the carried frame, thus independent of the system under consideration. With regard to the dynamical aspect, the problem can be reduced to that on the trajectory Γ, on condition that this supplementary kinetic momentum is added to the orbital momentum. This supplementary momentum thus appears as a specific property, necessary to the complete characterization of the particle, in addition to its mass and its charge, when one remains at the level of Schrödinger's formalism which, as we have seen in Chapter VIII, can be found by considering the electric part of the universe field only. The spin or the magnetic momentum of the electron have also to be introduced in the semi-classical problems where magnetic fields are involved (e.g. in the Stern-Gerlach experiment, *vide infra*) , and in the nuclear reactions (e.g. in the fission processes) where the total kinetic momentum must to be conserved.

Owing to the isotropy of the universe field, the intrinsic kinetic momentum can be written as follows

$$\sigma = N. f(t) \tag{IX-12}$$

N being a unitary vector whose direction varies randomly versus time, and $f(t)$ a *positive* function of time, uncorrelated with the direction of the vector N .

Another consequence of the isotropy of the universe field, is that the component of σ along any direction z is equal to zero. Indeed,

$$\overline{\sigma_z} = \overline{f(t)}.\overline{\cos\theta} = 0 \qquad\qquad \theta = (\sigma, z) \tag{IX-13}$$

On the other hand

$$\overline{\sigma_z^2} = \frac{1}{3}\overline{f^2} \qquad \text{and} \qquad \overline{\sigma^2} = \overline{f^2} \tag{IX-14}$$

Thus, according to (IX-11)

$$\overline{f^2} = \frac{3}{4}\hbar^2 \tag{IX-15}$$

and, according to the inequality $\overline{f^2} > (\overline{f})^2$

$$\overline{f} < \sqrt{3}\frac{\hbar}{2} \tag{IX-16}$$

Intrinsic magnetic momentum of electron

According to classical electrodynamics, to any particle of mass m and charge q, exhibiting a kinetic momentum $\mathbf{M}$, there is associated a magnetic momentum (we will call it classical) , proportional to $\mathbf{M}$

$$\mu_{class.} = -\frac{q}{mc}\mathbf{M} \qquad\qquad (IX\text{-}17)$$

This result should apply to the intrinsic kinetic momentum of the electron. But, as we have seen, the value of the magnetic momentum is twice the expected value. In his quantum theory, Dirac effectively accounts for this factor 2.

Our model allows to find this factor again as follows. The electron velocity on the trajectory Z being of the order of that of light, the situation has to be compared with that encountred in the betatron. The reader will find the theory of this device for instance in Ref [7] . The result which interests us for our problem, is that, for a given variation of the flow through a circular current created by relativistic electrons, the variation of the induction is twice that given by the classical theory, so that the magnetic energy accepted by the circuit when the external field varies from zero to $\mathbf{H}$, is twice that foreseen by the non-relativistic theory. From which it results the factor $g = 2$. Consequently

$$\mu = -\frac{q}{mc}\sigma \qquad\qquad (IX\text{-}18)$$

In fact, owing to the terms of second order with respect to the field, neglected in the expression of the damping force, the value of σ is slightly modified, so that, if, according to custom, we refer to the conventional spin-momentum, the factor g is very slightly different from 2.

Magnetic momentum of positron

Let us consider a positive electron or, more generally the antiparticle of mass m and charge $-q$ associated to the particle (m, q) . Given that the bursts received from the universe field are uncorrelated, over a sufficiently long time, the momentum σ is the sum of the contributions $\delta\sigma$ brought by the successive bursts. Now, owing to the isotropy of the field, to any burst (**E, H**), it corresponds the burst (-**E**, -**H**), so that every thing occurs as if

$$\mathbf{H}_{-q}(t) = -\mathbf{H}_q(t) \; ; \; \mathbf{E}_{-q}(t) = -\mathbf{E}_q(t) \tag{IX-19}$$

For such pairs of bursts, the equations of the motion of the particle and of its corresponding antiparticle are

$$-\tau\ddot{\rho}_q + \ddot{\rho}_q = \frac{q}{mc}\left[\dot{\rho}_q \wedge \mathbf{H}_q + \left(\rho_q\right)_n \dot{\mathbf{E}}_q\right] \tag{IX-20}$$

and

$$\begin{aligned}
-\tau\ddot{\rho}_{-q} + \ddot{\rho}_{-q} &= -\frac{q}{mc}\left[\dot{\rho}_{-q} \wedge \mathbf{H}_{-q} + \left(\rho_{-q}\right)_n \dot{\mathbf{E}}_{-q}\right] \\
&= \frac{q}{mc}\left[\dot{\rho}_{-q} \wedge \mathbf{H}_q + \left(\rho_{-q}\right)_n \dot{\mathbf{E}}_q\right]
\end{aligned} \tag{IX-21}$$

From which it results that

$$\rho_{-q}(t) = \rho_q(t) \tag{IX-22}$$

Consequently, for each pair of bursts satisfying the condition (IX-19), the trajectories are the same, so that the contributions to the kinetic momentum are equal

$$\delta\sigma_{-q}(-\mathbf{H},-\mathbf{E}) = \delta\sigma_q(\mathbf{H},\mathbf{E}) \tag{IX-23}$$

Thus, over a sufficiently long time, the kinetic momenta are equal and, according to (IX-18), the corresponding magnetic momenta are opposite.

The Vaschy theorem

According to this (not very known) theorem, the mutual energy between an ideal permanent magnet and an electric circuit is equal to zero. Let us recall that a permanent magnet is a magnet whose magnetization intensity is invariable whatever the value and the variations of the electromagnetic field to which it can be submitted. The actual magnets, of course, answer only very approximatively to this definition .

Anyhow, if, according to the ideas of Ampère, magnetism is due to currents of particles and not to magnetic charges (Coulomb model) , a magnet must behave as a current. Now the mutual energy of two circuits is different from zero. It then follows an inconsistency in which Louis de Broglie saw the proof of the existence of spin which confers a magnetic momentum to electron, independent of the exterior electromagnetic conditions [8] . A recent discussion [9] shows that, in fact, the Ampère model does not disagree with the Vaschy theorem.

In our model, the magnetic momentum of the electron arises from its motion on the trajectory Z. Hence, its origin is a current. But the motion on Z is independent of the variations of the induction flow through Z. The flow of an exterior field is, indeed, equal to the circulation of the corresponding vector potential along Z, $\overline{\mathbf{A}_{ext}\dot{\rho}}$, i.e. $\overline{\mathbf{A}_{ext}.\dot{\rho}}$ which is equal to zero given the ergodicity of the motion. Consequently, our model is not inconsistent with the Vaschy theorem.

The situation is more complex for an actual magnet. Indeed, the ferromagnetism of a material is interpreted by assuming that this material is constituted by small domains exhibiting an uniform magnetization (the Weiss domains) created by a rigid set of parallel electron spins. But the sizes and the orientations of these domains can vary under the effect of the external fields, which explains the difference between the ideal magnet considered by Vaschy and the real ones.

In any case, it is interesting to remark that, in order to interpret the macroscopic magnetism, it is not necessary to introduce magnetic point charges. Our model would rather militate against the existence of the latter.

Effect of a constant magnetic field

Let us assume that, independently of the universe field, the system is immersed within a constant magnetic field $\mathbf{H}_0$. As we have seen in

Chapter VII, the trajectory Γ is unchanged. The effect of $\mathbf{H}_0$ appears only at the level of the trajectory Z. In particular, the intrinsic kinetic momentum of the electron will be modified.

According to Eq (IX-5) , the equation which defines the new trajectory is

$$-\tau\ddot{\rho}+\ddot{\rho}=\frac{q}{m}\frac{\dot{\rho}}{c}\wedge(\mathbf{H}+\mathbf{H}_0)-\frac{q}{m}\frac{\dot{\rho}_n}{c}\,\mathbf{E} \qquad \text{(IX-24)}$$

In order to make the kinetic momentum appear, let us multiply this equation by $m\rho\wedge$ .We obtain

$$-\tau(\rho\wedge m\ddot{\rho})+(\rho\wedge m\ddot{\rho})=q\rho\wedge\left[\frac{\dot{\rho}}{c}\wedge(\mathbf{H}+\mathbf{H}_0)\right]-q\frac{\dot{\rho}}{c}\rho\wedge\mathbf{E}$$

i.e.

$$-\tau\ddot{\sigma}+\tau(\dot{\rho}\wedge m\ddot{\rho})+\dot{\sigma}=q\rho\wedge\left[\frac{\dot{\rho}}{c}\wedge(\mathbf{H}+\mathbf{H}_0)\right]-q\frac{\dot{\rho}_n}{c}\rho\wedge\mathbf{E} \qquad \text{(IX-25)}$$

Now for any field, we can successively write

$$\begin{cases}\rho\wedge(\dot{\rho}\wedge\mathbf{H})=(\rho\mathbf{H})\dot{\rho}-(\rho\dot{\rho})\mathbf{H}\\ \mathbf{H}\wedge(\rho\wedge\dot{\rho})=(\mathbf{H}\dot{\rho})\rho-(\mathbf{H}\rho)\dot{\rho}\end{cases} \qquad \text{(IX-26)}$$

Hence

$$\rho\wedge[\dot{\rho}\wedge(\mathbf{H}+\mathbf{H}_0)]=\frac{\sigma}{mc}\wedge(\mathbf{H}+\mathbf{H}_0)-(\rho\dot{\rho})(\mathbf{H}+\mathbf{H}_0)-[\rho(\mathbf{H}+\mathbf{H}_0)]\dot{\rho}$$

$$\text{(IX-27)}$$

i.e., for the equation of the motion

$$-\tau\ddot{\sigma}+\dot{\sigma}+\tau(\dot{\rho}\wedge m\ddot{\rho})=\frac{q}{mc}\sigma\wedge(\mathbf{H}+\mathbf{H}_0)-q\left(\frac{\dot{\rho}}{c}\rho\right)(\mathbf{H}+\mathbf{H}_0)$$

$$-q[\rho(\mathbf{H}+\mathbf{H}_0)]\frac{\dot{\rho}}{c}-q\frac{\rho_n}{c}\dot{\rho}\wedge\mathbf{E} \qquad \text{(IX-28)}$$

A priori, the direction of $\mathbf{H}$ varies much more quickly than that of σ so that we can replace $(\mathbf{H}+\mathbf{H}_0)$ by its average value $\mathbf{H}_0$. On the other hand, the term in which the electric field appears, is of the same order of magnitude as the neglected magnetic terms. Finally, we can replace the product $\dot{\rho}\wedge m\ddot{\rho}$ by its average value on Z which is equal to zero. Consequently, we obtain the following equation

$$-\tau\ddot{\sigma} + \dot{\sigma} = \frac{q}{mc}\sigma \wedge \mathbf{H}_0 \qquad \text{(IX-29)}$$

which is the equation corresponding to a magnetic dipole σ within the magnetic field $\mathbf{H}_0$ when one takes its radiation into account [10].

A rather long but straightforward calculation allows to lead to the following differential equation

$$-\tau\ddot{\theta} + \dot{\theta}\left(1 - 2\tau\frac{\dot{f}}{f}\right) + \tau\omega^2 \sin\theta \cos\theta = 0 \qquad \text{(IX-30)}$$

where θ is the angle $(\sigma, \mathbf{H}_0)$, f the modulus of σ and ω the Larmor frequency $q|\mathbf{H}_0| / mc$.

Practically, the term $\tau\ddot{\theta}$ can be neglected. Indeed, if we replace the term $(1 - 2\tau\dot{f}/f)$ by its average value a (which is possible because f varies much more quickly than the direction of σ), and if we neglect this term, the equation reduces to $\dot{\theta} = -\omega^2\tau\sin 2\theta / 2a$ i.e. $\ddot{\theta} = -\omega^2 \cos 2\theta.\dot{\theta}$.

Thus, $|\tau\ddot{\theta}| \approx \frac{\omega^2\tau^2}{a^2}|a\dot{\theta}|$ which shows that $|\tau\ddot{\theta}| << |a\dot{\theta}| \approx \left|\left(1 - 2\tau\frac{\dot{f}}{f}\right)\dot{\theta}\right|$ if $\omega\tau << a$. Now for a field of 10^4 gauss, as in the Stern and Gerlach experiment, $\omega \cong 10^{11}s^{-1}$, so that the condition is satisfied, and the solution of (IX-30) is

$$\tan\theta = \tan\theta_0 \exp\left[-\omega^2\tau\int_0^t \frac{dt}{1 - 2\tau\dot{f}/f}\right] \qquad \text{(IX-31)}$$

θ_0 being the angle which the momentum σ makes with the field $\mathbf{H}_0$ when the atom which carries the electron under consideration enters into the field of the magnet [11].

If the modulus of σ, namely f, is constant, we have

$$\tan\theta = \tan\theta_0 \exp(-\omega^2\tau t) \qquad \text{(IX-32)}$$

Consequently, if $0 < \theta_0 < \pi/2$, σ orients in a parallel manner with respect to the field $(\theta \to 0)$, and if $\pi/2 < \theta_0 < \pi$, in an antiparallel manner $(\theta \to \pi)$. We find again the alignment effect of the spin momentum with respect to the field according to the orientation of this momentum when it enters into the field. Nevertheless, the result is quantitalively unacceptable. Indeed, within a 10^4 gauss field $(\omega \cong 10^{11} s^{-1})$, the time necessary to pass from $\theta_0 = 89°$ to $\theta = 1°$ would be of the order of one minute whereas the flight-time of the atom in the device is of the order of $10^{-3} s$. This result, previously obtained by the authors of Ref [10], was interpreted by them as being the proof that the Stern and Gerlach experiment cannot be classically explained. In our model, this result is not surprising because f does vary.

In order to make the orientation effect precise, we will assume that f varies periodically according to the law

$$f \propto \sin^2 \alpha t$$

over a burst of a $\pi/2\alpha$ life, falling back to zero just before the following burst. The number of bursts received from the universe field being very great over the time τ, the product $\alpha\tau$ is very superior to 1, so that the integral which appears in Eq (IX-31) is equal to

$$\int_0^t \frac{\sin \alpha t}{-4\alpha\tau \cos \alpha t + \sin \alpha t} dt$$

$$= \frac{1}{\alpha(1+16\alpha^2\tau^2)} \left[-4\alpha^2\tau t + \ln \frac{\sin arctan(4\alpha\tau)}{\sin[arctan(4\alpha\tau) - \alpha t]} \right] \qquad (IX\text{-}34)$$

i.e., by remarking that $\sin arctan(4\alpha\tau) \cong \sin \frac{\pi}{2} = 1$

$$-\frac{1}{16\alpha^3\tau^2} \left[4\alpha^2\tau t + \ln \cos \alpha t \right] \qquad (IX\text{-}35)$$

which tends to $+\infty$ when t tends to $\pi/2\alpha$, so that $\tan\theta$ tends to zero.

This calculation, of course, is only a model calculation, but, nevertheless, it shows that the sudden variations of the modulus of the kinetic momentum is orientating the latter along the magnetic field in a practically instantaneous time. This is an example of a classical interpretation of what quantum mechanics calls the reduction of the wave-packet, here the set of the two components $s_z = \pm\hbar/2$ into one of these components $+\hbar/2$ or $-\hbar/2$.

Under the effect of the field $\mathbf{H}_0$ the average value of the component of the kinetic momentum along the direction of this field is equal to $\pm\bar{f}$. Now experiment shows that this component is equal to $\pm\hbar/2$, so that

$$\bar{f} = \frac{\hbar}{2} \tag{IX-36}$$

which is quite compatible with relationship (IX-16).

Correlation in a singlet state

Although we have up to now considered single particle problems only, we can now examine the question of the correlation between the measurements of the spin momenta of two electrons in a singlet state, along the directions z and z' respectively, making an angle ω between them.

Quantum mechanics obtains

$$C_{zz'} = \langle s_z(1)s_{z'}(2)\rangle = -\cos\omega \tag{IX-37}$$

for the correlation coefficient, the values of the s_z's being normalized to unit.

Let $\mathbf{u}$ (1,0,0) and $\mathbf{u}'($ $\cos\omega$, $\sin\omega$, 0) be the unitary vectors corresponding to the directions z and z' respectively. The directions of the vectors σ_1 and σ_2 corresponding to the electrons 1 and 2 respectively are defined by their respective components

$$\begin{array}{lll} \sin\theta\cos\varphi, & \sin\theta\sin\varphi, & \cos\theta \\ \sin\theta'\cos\varphi', & \sin\theta'\sin\varphi', & \cos\theta' \end{array} \tag{IX-38}$$

(θ, φ corresponding to electron 1, and θ', φ' to electron 2).

The measurement along z for electron 1 gives $f_1 \sin\theta\cos\varphi$ and along z' for electron 2 $f_2(\cos\omega\sin\theta'\cos\varphi' + \sin\omega\sin\theta'\sin\varphi')$.

In order to obtain a correlation coefficient which is able to be compared with that given by quantum mechanics (IX-37), we must normalized the measurement results. The square of the norm is the same for the two electrons. It is equal to

$$\overline{f^2 . \sin^2 \theta . \cos^2 \varphi} = \frac{1}{3}\overline{f^2} = (\tilde{s})^2 \qquad \text{(IX-39)}$$

From which one gets

$$C'_{zz'} = \frac{3\overline{f_1 f_2}}{f^2}\overline{\left[\sin\theta\sin\theta'\left(\cos\varphi\cos\varphi'\cos\omega + \cos\varphi\sin\varphi'\sin\omega\right)\right]} \quad \text{(IX-40)}$$

In order to calculate this average value, we must take into account the fact that the four variables θ, θ', φ, φ' are not independent of one another. The latter, indeed, are connected with the angle α made by the vectors σ_1 and σ_2 by the following relationship

$$\sin\theta\sin\theta'\cos(\varphi'-\varphi) + \cos\theta\cos\theta' = \cos\alpha \qquad \text{(IX-41)}$$

Thus

$$\cos(\varphi'-\varphi) = \frac{\cos\alpha - \cos\theta\cos\theta'}{\sin\theta\sin\theta'}$$

$$\sin(\varphi'-\varphi) = \pm\frac{\sqrt{\sin^2\theta\sin^2\theta' - (\cos\alpha - \cos\theta\cos\theta')^2}}{\sin\theta\sin\theta'} \qquad \text{(IX-42)}$$

The variables $\theta, \theta', \varphi, \alpha$ are independent of one another. Hence

$$C'_{zz'} = \frac{\overline{f_1 f_2}}{f^2}\overline{\cos\alpha . \cos\omega} \qquad \text{(IX-43)}$$

We find the quantum result again if, and only if, σ_1 and σ_2 are *opposed at any time* ($f_1 = f_2$ and $\alpha=\pi$).

Such a situation which is not surprising in quantum mechanics, is worth being examined in our model. We have seen that in their motions on their respective trajectories Γ each of the electrons carries an intrinsic kinetic momentum corresponding to their trajectories Z. Let $\sigma_1 = \rho_1 \wedge m\dot{\rho}_1$ and $\sigma_2 = (r_{12} + \rho_2) \wedge m\dot{\rho}_2$ be these two momenta at a given time (the term r_{12} has been introduced in order that σ_1 and σ_2 be referred to the same origin of the coordinates) . Under the effect of a burst, the variations of these momenta will be respectively

$$\delta\sigma_1 = \delta\rho_1 \wedge m\dot\rho_1 + \rho_1 \wedge m\delta\dot\rho_1$$

$$\delta\sigma_2 = (\delta\mathbf{r}_{12} + \delta\rho_2) \wedge m\dot\rho_2 + \rho_2 \wedge m\delta\dot\rho_2 + \mathbf{r}_{12} \wedge m\delta\dot\rho_2 \qquad \text{(IX-44)}$$

$$\approx \delta\rho_2 \wedge m\delta\dot\rho_2 + \rho_2 \wedge m\delta\dot\rho_2$$

by neglecting the variation of $\mathbf{r}_{12}$ during the burst and the term $\mathbf{r}_{12} \wedge m\delta\dot\rho_2$ whose time-average value is equal to zero.

Now, for each particle, the variation of the velocity is $\delta\dot\rho_H$. This variation is of the same sign as that of $\dot\rho$ whereas the variation of the position is independent of this sign. Consequently, the sign of $\delta\sigma$ is that of the initial velocity.

Let us assume that, at a given time, the two momenta are antiparallel with opposite speeds. According to the foregoing, if we neglect the time necessary for the perturbation to go from a particle to the other, the two momenta will vary, remaining opposite. Likewise, if the momenta are parallel with the same speed, they remain parallel. Thus, starting from an arbitrary relative disposition, under the effect of chance, after a sufficiently great number of bursts, the momenta will become parallel or antiparallel, and then they will remain trapped in the configuration reached.

Let us examine the propagation effect we have neglected in a first approach. The previous result has to be stated more precisely as follows: The value of σ_1 at the time t is equal to that of σ_2 or $-\sigma_2$ at the time $t - \tau_{21}$: $\sigma_1(t) = \pm\sigma_2(t - \tau_{21})$, τ_{21} being the time necessary for the perturbation to go from the point where electron 2 is located to that where electron 1 is. In an atom, the distances being typically of about $10^{-8}\,cm$, τ_{21} is of the order of $10^{-18}\,s$, which is very great compared with the time interval between two successive bursts. Now

$$\sigma_1(t) = \varepsilon\sigma_2(t - \tau_{21}) = \varepsilon\sigma_2(t) - \varepsilon\tau_{21}\dot\sigma_2(t) + \dots \qquad \text{(IX-45)}$$

with $\varepsilon = \pm 1$, so that in carrying out the averaging over a time interval T able to be experimentally reached, i.e. much greater than τ_{21}, and, a fortiori, than the interval between two successive bursts, we obtain

$$\overline{(\sigma_1 - \sigma_2)_T} = -\frac{\tau_{21}}{T}[\varepsilon\sigma_2]_0^T \cong 0 \qquad \text{(IX-46)}$$

which signifies that, for an observer, the momenta are either parallel, or antiparallel. The difference will be observable for macroscopic distances only, for instance for the two electrons of a molecule H_2 ($\varepsilon = -1$) when the internuclear distance tends to infinity. Thus, during the dissociation of the molecule, as the distance increases, the spin momenta of the two electrons will be less and less correlated with one another, so that, at the limit, we obtain two independent hydrogen atoms. This conclusion is, of course, more satisfying than that deduced from the orthodox quantum mechanics according to which the complete correlation between the two electrons is preserved even for arbitrary large distances, which would involve that the two atoms would kept the memory of their past.

The Bell inequality

Certainly, after the epic dispute between Bohr and Einstein about the interpretation of quantum mechanics, the Bell work [12] marked this theory in a deeper manner than all the other ones. This author, indeed, showed that a hidden variable model is inconsistent with the quantum mechanics.

Let us briefly recall the Bell proof. Owing to the definition of the correlation coefficient, for the directions a and b, we have

$$C_{ab} = \overline{s_a(1)s_b(2)} \qquad \text{(IX-47)}$$

with $s_a(1)$ and $s_b(2)$ equal to ± 1, the selection of the sign depending on a hidden variable.

Let us introduce a third direction c. We obtain

$$C_{ab} - C_{ac} = \overline{s_a(1)s_b(2)} - \overline{s_a(1)s_c(2)}$$

$$= \overline{s_a(1)s_b(2)\left[1 - \frac{s_b(2)s_c(2)}{s_b{}^2(2)}\right]} \tag{IX-48}$$

i.e., by using relationship $\overline{|xy|} \le |x|_{max} \cdot \overline{|y|}$ and the fact that $s_b{}^2 = 1$,

$$|C_{ab} - C_{ac}| \le |s_a(1)s_b(2)|_{max} \cdot \overline{|1 - s_b(2)s_c(2)|}$$
$$= 1 - \overline{s_b(2)s_c(2)} \tag{IX-49}$$

Hence, for a singlet state for which $s_b(2) = -s_b(1)$

$$|C_{ab} - C_{ac}| \le 1 + C_{bc} \tag{IX-50}$$

i.e. according to (IX-37)

$$|\cos(a,b) - \cos(a,c)| \le 1 - \cos(b,c) \tag{IX-51}$$

This relationship is violated for certain directions. For instance if $(a,b) = \pi/2$ and $(a,c) = \pi/4$, the left-member of (IX-51) is equal to $1/\sqrt{2}$ whereas the right-member is equal to $1 - 1/\sqrt{2}$.

According to the usually admitted opinion, the violating of inequality (IX-51) has to be imputed to the *locality* hypothesis, implicitly made from the beginning, and according to which the result obtained for electron 1 along the direction a depends neither on the direction along which the measurement for electron 2 is performed, nor on the result of this measurement. But, to reject this well natural hypothesis amounts to saying that an interaction does exist between the two devices which measure $s_a(1)$ and $s_b(2)$, or also between the electrons themselves whatever the distances which separate them from each another. In other words, such an interpretation signifies that it exist signals able to travel with a velocity greater than that of light, which questions the whole relativity theory!

Let us see how the problem can be resolved in our model, independently of the fact that the correlation coefficients C_{xy} tend to zero for macroscopic distances.

The spin varying both in modulus and in direction, in order to reduce the problem to that of a variable able to exhibit the values +1 and -1 only, we will adopt +1 if the value obtained for the component under consideration is positive, and -1, if the component is negative. Such a dichotomy amounts to considering the sign of the projection of the spin along the corresponding directions only. As it is easy to see, Eq (IX-37) has to be replaced by the relationship

$$C_{ab} = -1 + \frac{2}{\pi}(a,b) \qquad\qquad \text{(IX-52)}$$

where (a,b) is the angle ($\leq \pi$) of the positive directions a and b . We will notice the difference in the treatment of the information with that which has lead us to formula (IX-50) . The two relationships (IX-37) and (IX-52) give, nevertheless, the same results for 0, $\pi/2$ and π .

Thus, according to (IX-50), we obtain to the never violated inequality

$$|(a,b) - (a,c)| \leq (b,c) \qquad\qquad \text{(IX-53)}$$

Consequently, the violation of the Bell inequality [13] , if indeed it is significant [14] , and the conclusions which can be derived from it [15] , arise from a too restrictive hypothesis made at the beginning concerning the result of the measurement of the spin components, and not from any physical reality. This conclusion is a supplementary argument for our model. In passing, we will notice that we join again the idea of Einstein [16] and that of Louis de Broglie [17] which categorically refused to abandon the local character of physics.

References

[1] O. Stern, Z. *Phys.* , **7** (1921) 249 ; W. Gerlach, O. Stern, *ibidem,* **8** (1922) 110 ; **9** (1922) 353.

[2] G. Goudsmit, S. Uhlenbeck, *Naturwissenschaften* , **13** (1925) 953.

[3] H. Corben, *Classical and Quantum Theories of Spinning Particles* (Holden-Day, San-Fransisco) 1968 ; D. Hestenes, *J. Math. Phys.* **14** (1973) 893 ; A.O. Barut, A.J. Brocken, *Phys. Rev. D*, **23** (1981) 2454 ; **24** (1981) 3333 ; M. Rivas, *J. Math. Phys.*, **30** (1989) 318.

[4] L. de la Peña, A. Jáuregui, *Found. Phys.*, **12** (1982) 441.

[5] A. Julg, *J. Molecular Structure (Theochem)* **330** (1995) 85.

[6] D. Hestenes, *Am. J. Phys.*, **47** (1979) 399.

[7] Y. Rocard, *Electricité* (Masson, Paris) 1951, p. 362.

[8] L. de Broglie, *Portugaliae Physica*, **31** (1949) 1.

[9] O. Costa de Beauregard, *Ann. Fondation L. de Broglie*, **17** (1992) 199.

[10] A. Rañada, F. Rañada, *J. Phys. A,* **12** (1979) 1419.

[11] A. Julg, *Ann. Fondation L. de Broglie*, **16** (1992) 321.

[12] J. Bell, *Physics*, **1** (1965) 195.

[13] S.J. Freedman, J.F. Clauser, *Phys. Rev. Lett.*, **28** (1972) 938 ; J. F. Clauser, *Phys. Rev. Lett.*, **37** (1976) 1223 ; A. Aspect, P. Grangier, G. Roger, *Phys. Rev. Lett.*, **47** (1981) 460 and **49** (1982) 91 ; W. Perrie, A.J. Duncan, H.J. Beyer, H. Kleinpoppen, *Phys. Rev. Lett.*, **54** (1985) 1790 ; D. Bouwmeester, J.W. Pan, K. Hattle, M. Eibl, H. Weinfinter, A. Zeilinger, *Nature* **390** (1997) 575 (Quantum teleportation) .

[14] V.L. Lepore, F. Selleri, *Foundations of Physics Letters,* **3** (1990) 203.

[15] For instance, see : B. d'Espagnat, *Une incertaine réalité* (Gauthier-Villars, Paris) 1985 ;
C.H. Bennet, G. Brassard, C. Crepan, R. Jozsa, A. Peres, W. Wooters, *Phys. Rev. Lett.* **70** (1993) 1895 (Quantum teleportation theory) .

[16] A. Einstein, N. Rosen, B. Podolsky, *Phys. Rev.*, **47** (1935) 777 ; A. Einstein in *Albert Einstein, Philosopher Scientist*, P.A. Schilp ed (Library of living philosophers, Evanston, Ill.) 1949, p. 85.

[17] L. de Broglie in L. de Broglie, J.L. Andrade e Silva, *La réinterprétation de la mécanique ondulatoire* . 1 *Principes généraux* (Gauthier-Villars, Paris) 1971, p. 93.

X

The excited states

The quantal point of view

The problem of the excited states is probably the domain in which theoreticians proclaim the superiority of quantum mechanics with the greatest vigor, asserting that the latter is the sole theory able to account for the existence of discrete spectra. In the classical physics, indeed, the energy of any system, the hydrogen atom for instance, can exhibit an arbitrarily fixed value inside a large range. Under these conditions, one does not see how a discrete spectrum could be explained.

On the contrary, all seems simple in the quantal outlook. According to Schrödinger's equation, indeed, the system can exhibit well-determined energies only, namely the eigenvalues of the time-independent Hamiltonian operator. The time-dependent perturbation theory shows that a radiation can interact with the system - thus to be absorbed - only if its frequency ν is such as

$$\Delta E = h \nu \qquad \text{(X-1)}$$

ΔE being the difference between one of the eigenvalues of the Hamiltonian and that corresponding to the initial state. From which , the idea of *jumps* between pre-existent levels, true shelves on which electrons can take place. The time-independent perturbation theory, by its language, had reinforced such an image.

In fact, the situation is far from being clear. The strict application of the Schrödinger formalism accounts neither for these jumps, nor for the back transitions to the initial state when the light excitation is switched off [1]. And, what is worse, the calculation [2] shows that the system does *oscillate* between the initial state and that foreseen by the conventional theory, with a frequency of about the millionth of that of the radiation !

Briefly, let us demonstrate this unfortunately not very known result.

In this view, we will consider a 2-state system. Let $\Psi_p^0(t) = \psi_p(x,y,z)\exp(i\omega_p t)$ ($p = 1, 2$) be the respective time-dependent wave-functions of the two states, and $E_p = \hbar\omega_p$ the correspoding energies. At the time $t = 0$, let us assume that the system is immersed in an electromagnetic radiation of frequency ω, deriving from the following vector potential $\mathbf{A}(t)$ $\left\{A_x = A_0\cos\omega t, A_y = 0, A_z = 0\right\}$. Then the system is governed by the following equation

$$i\hbar\frac{\partial\Psi}{\partial t} = \left(\hat{H}_0 + \hat{V}\right)\Psi \tag{X-2}$$

$\hat{H}_0$ being the hamiltonian in the absence of the radiation, and $\hat{V}$ the operator $\dfrac{i\hbar e}{mc}A_0\dfrac{\partial}{\partial x}$. The solution is of the following form

$$\Psi_1(t) = \sum_{k=1,2} a_k(t)\Psi_k^0(t) \tag{X-3}$$

with

$$a_1(0) = 1 \quad \text{and} \quad a_2(0) = 0 \tag{X-4}$$

By substituting $\Psi_1(t)$ (X-3) into (X-2) , and after integrating with respect to the space variables, we obtain

$$\begin{cases} i\hbar\dot{a}_1 = a_2 V_{12} \\ i\hbar\dot{a}_2 = a_1 V_{21} \end{cases} \tag{X-5}$$

with $V_{pq} = \dfrac{\omega_p - \omega_q}{ic}QA_0\cos\omega t\exp i(\omega_q - \omega_p)t$ \qquad (X-6) where $Q = \left| \langle \psi_p e \mathbf{R} \psi_q \rangle \right|$ is the value of the moment of the transition $p \to q$ (or $q \to p$).

Let us assume that a_1 and a_2 vary slowly with respect to the terms in V which depend on ω. Then we are in the case of systems governed by two classes of variables, the ones varying quickly and the others varying slowly [3]. In these cases it is possible, in a first step, to replace V_{12} and V_{21} in Eq(X-5) by their respective average value and to neglect the derivatives of these terms whose respective time average values are equal to zero, as it is easy to verify from (X-6).

Consequently, Eq(X-5) reduces to
$$\begin{cases} i\hbar \dot{a}_1 = a_2 \overline{V_{12}} \\ i\hbar \dot{a}_2 = a_1 \overline{V_{21}} \end{cases} \tag{X-7}$$

Now, when ω tends to $(\omega_2 - \omega_1)$
$$\overline{V_{12}} = \overline{V_{21}}* = \frac{i\omega Q A_0}{2c} \tag{X-8}$$

Therefore, we have
$$\begin{cases} \ddot{a}_1 + \Omega^2 a_1 = 0 \\ \ddot{a}_2 + \Omega^2 a_2 = 0 \end{cases} \tag{X-9}$$

with
$$\Omega = \frac{\omega Q A_0}{2\hbar c} \cong 10^{-6}\,\omega \tag{X-10}$$

which justifies the approximation we have made.

Then
$$\begin{cases} a_1 = \cos \Omega t \\ a_2 = -\sin \Omega t \end{cases} \tag{X-11}$$

The true solution of (X-5) will be obtained by quickly superimposing variations of small amplitude arising from the terms in ωt in the terms V, on the solution (X-11) without the oscillating behavior of the solution being modified.

In reality, this paradoxical result arises simply from the fact that absorption is a phenomenon too quick with respect to the ergodicity time of the electron systems for the time-dependent Schrödinger equation to be applicable (See Chap. VIII) .

The excited state in our model

Up to now, we have considered the system as being immersed within the universe field only. In this case, we have seen that the system reaches a well-determined equilibrium, corresponding to the so-called *ground state* . Let us now assume that, in addition to the universe field, the system is immersed within the electromagnetic field of a radiation of frequency ω .

In the absence of this radiation, the particle described a certain trajectory $\Gamma(\mathbf{R})$ defined by Eq. (VII-6)

$$-m\tau\dddot{\mathbf{R}} + m\ddot{\mathbf{R}} = \mathbf{F}(\mathbf{R}) + q\mathbf{E} \tag{X-12}$$

which, by multiplying by $\dot{\mathbf{R}}$, gave the following average values (VII-25)

$$m\tau\overline{\dddot{\mathbf{R}}^2} = q\overline{\dot{\mathbf{R}}\mathbf{E}} \tag{X-13}$$

In the presence of the radiation, we obtain a new trajectory $\Gamma^*(\mathbf{R}^*)$ with $\mathbf{R}^* = \mathbf{R} + \mathbf{r}$, defined by

$$-m\tau(\dddot{\mathbf{R}} + \dddot{\mathbf{r}}) + m(\ddot{\mathbf{R}} + \ddot{\mathbf{r}}) = \mathbf{F}(\mathbf{R} + \mathbf{r}) + q(\mathbf{E} + \mathbf{E}_{rad}) \tag{X-14}$$

$\mathbf{E}_{rad}$ being the electric field of the radiation.

Let us multiply this equation by $(\dot{\mathbf{R}} + \dot{\mathbf{r}})$ and integrate over a sufficiently long time. We obtain

$$m\tau\overline{(\dddot{\mathbf{R}} + \dddot{\mathbf{r}})^2} = q\overline{(\dot{\mathbf{R}} + \dot{\mathbf{r}})(\mathbf{E} + \mathbf{E}_{rad})} \tag{X-15}$$

which generalizes (X-13) .

Now $\mathbf{R}$ depends on $\mathbf{E}$, and $\mathbf{r}$ on $\mathbf{E}$ and $\mathbf{E}_{rad}$. $\mathbf{E}$ and $\mathbf{E}_{rad}$ being non-correlated, the average value of $\dot{\mathbf{R}}\mathbf{E}_{rad}$ is equal to zero. Thus Eq(X-15) reduces to

$$m\tau\overline{(\dddot{\mathbf{R}} + \dddot{\mathbf{r}})^2} = q\overline{(\dot{\mathbf{R}} + \dot{\mathbf{r}})\mathbf{E}} + q\overline{\dot{\mathbf{r}}\mathbf{E}_{rad}} \tag{X-16}$$

i.e., according to (X-13)

$$m\tau(2\overline{\dddot{\mathbf{R}}\dddot{\mathbf{r}}} + \overline{\dddot{\mathbf{r}}^2}) = q\overline{\dot{\mathbf{r}}\mathbf{E}} + q\overline{\dot{\mathbf{r}}\mathbf{E}_{rad}} \tag{X-17}$$

By difference between (X-14) and (X-12), after multiplication by $\dot{\mathbf{r}}$, we obtain

$$m\tau\overline{\dot{\ddot{\mathbf{r}}}^2} = q\overline{\dot{\mathbf{r}}\mathbf{E}_{rad}} \tag{X-18}$$

So that we have

$$2m\tau\overline{\dot{\mathbf{R}}\ddot{\mathbf{r}}} = q\overline{\dot{\mathbf{r}}\mathbf{E}} \tag{X-19}$$

On the other hand, if we multiply (X-14) by $\dot{\mathbf{r}}$, we obtain

$$m\tau\overline{\ddot{\mathbf{r}}^2} = q\overline{\dot{\mathbf{r}}\mathbf{E}} + q\overline{\dot{\mathbf{r}}\mathbf{E}_{rad}} \tag{X-20}$$

Consequently

$$\overline{\dot{\mathbf{r}}\mathbf{E}} = \overline{\dot{\mathbf{r}}\dot{\mathbf{R}}} = 0 \tag{X-21}$$

On the trajectory Γ the average radiated and absorbed powers are respectively

$$P_{rad} = m\tau\overline{\ddot{\mathbf{R}}^2} \quad \text{and} \quad P_{abs} = q\overline{\dot{\mathbf{R}}\mathbf{E}} \tag{X-22}$$

and on Γ^*

$$P^*_{rad} = m\tau\overline{(\ddot{\mathbf{R}} + \ddot{\mathbf{r}})^2} \quad \text{and} \quad P^*_{abs} = q\overline{(\dot{\mathbf{R}} + \dot{\mathbf{r}})(\mathbf{E} + \mathbf{E}_{rad})} \tag{X-23}$$

From these equations, it results

$$\begin{cases} P^*_{rad} = P_{rad} + m\tau\overline{\ddot{\mathbf{r}}^2} \\ P^*_{abs} = P_{abs} + q\overline{\dot{\mathbf{r}}\mathbf{E}_{rad}} \end{cases} \tag{X-24}$$

Thus, according to (X-18), on Γ^*, as on Γ, the average radiated power is equal to the absorbed one, which shows that, as for the ground state, we obtain an equilibrium state between the system and the rest of the universe.

Let us assume that the radiation is switched off. If the trajectory Γ^* remained unchanged, the average radiated power would be the same whereas the absorbed power would become equal to P_{abs}. Consequently, if the particle continued to follow the trajectory Γ^*, it would lose more energy than it would absorb. Thus, the state corresponding to the trajectory Γ^* is unstable when the radiation is suppressed.

Consequently, an excited state appears as a *provisional* situation due to the temporary modification of the exterior electromagnetic field, this situation ceasing as soon as the radiation is switched off. An excited state has not an own existence as in the orthodox quantum mechanics.

·We will notice, nevertheless, that the previous calculation does not give information concerning the time which is necessary for the system to be come back to the ground state.

In order to determine the characteristics of the excited state, we will, in first, prove an important proposition.

Theorem

The theory concerning the ground state introduces a certain constant K which appears in the product of the quadratic dispersions Δx, Δp_x, ... for a particle submitted to the universe field only. For a state corresponding to the superposition of this field and the one of a radiation, formally, the theory will be the same, but with another constant K^* which a priori will be different from K. Let us show that, in fact, K^* is equal to K, i.e. according to (VIII-21)

$$\hbar^* = \hbar \tag{X-25}$$

In this view, let us go back to Eq (VII-11) and (VII-12) which allowed us to reach the constant K, by adding the vector potential of the radiation $\mathbf{A} = \dfrac{c\mathbf{E_0}}{\omega} \cos \omega t$ to that of the universe field ($\mathbf{E}_{rad} = \mathbf{E_0} \sin \omega t$).

For the momentum, over a time interval τ, we obtain the following order of magnitude

$$\overline{(m\dot{x})^2} \approx \overline{(m\dot{x})_0^2} + \left(\frac{q\mathbf{E_0}\tau}{\omega}\right)^2 \tag{X-26}$$

(The index *zero* refers to the state in the absence of radiation)

For the most intense laser radiations which we can produce at the present time ($\cong 10^4 \, watt \, / \, cm^2$), $\mathbf{E_0}$ is of about 3000 V, i.e. $10^6 \, ues - cgs$. Consequently, even with a radiation hundred times more intense than the preceding one ($\mathbf{E_0} \approx 10^7$), for a visible wavelength - say 600 nm - the term $q\mathbf{E}_0\tau/\omega$ is of about $10^{-41} cgs$. Now we have seen (VII-18) that $\delta x \approx \tau \delta \dot{x}$, from which it results that $\overline{(m\dot{x})_0^2} \cong \dfrac{Km}{\tau} \cong 10^{-31} cgs$ so that the contribution of the radiation in Eq (X-16) is completely negligible compared to that arising from the universe field. Likewise for $\overline{x^2}$, hence the announced proposition (X-25).

Contrary to what occurs for $\hbar$, we will notice that the electromagnetic field of the radiation is acting on the arcs C of the trajectory Γ, i.e., finally, on the trajectory described by the particle. The trajectory Γ^* being different from Γ, the system will exhibit properties different from those of the ground state.

Consequences

The preceding theorem shows that the operator formalism which allows to obtain the various average properties (Chap. VIII) remains unchanged. In particular, the same hamiltonian operator will govern the ground state as well as the excited states. Now the average energies of the various states which are stable in the presence of the radiation, are stationary with respect to any variation of the trajectory (the proof is the same as for Γ). Consequently, as for the ground state, the average energies of the various excited states will be equal to the eigenvalues of the Hamiltonian. Thus, all the eigenvalues of the latter take a physical meaning.

In the presence of an electromagnetic radiation of frequency ω, the system passes from the ground state to a stationary state i whose average energy $\overline{E_i} = \overline{E_0} + \Delta E$ (X-27)

$\overline{E_i}$ being one of the eigenvalues of the Hamiltonian.

The Franck-Condon principle

The study of vibroelectronic spectra of molecules shows that the change of the electronic state occurs without modifying the positions and the speeds of the nuclei . This is the well-known Franck-Condon principle.

Let us assume that the source which emits the radiation yields a very usual power of $10^3 watt / cm^2$. The cross-section of an atom or a molecule is typically of the order of a_0^2 (a_0 = Bohr radius). We can thus estimate that the power absorbed by the system does not exceed $10^{-6} erg / s$. Given that the transition energies are of the order of eV, i.e. of $10^{-12} erg$, $10^{-6} s$ are necessary for such a radiation to be absorbed, which is inconsistent with the Franck-Condon principle owing to the fact the periods of the molecular vibrations are of about 10^{-11} - 10^{-12} s. One million periods of vibration would be necessary for such an energy to be absorbed. This is still an example of difficulties which the orthodox interpretation of quantum mechanics encounters when it refers to jumps between levels.

On the contrary, our model is consistent with this result while satisfying the Franck-Condon principle. Before the radiation is established, the electron describes a certain trajectory Γ . As soon as the radiation is established, the electron bifurcates towards another trajectory Γ^* which will correspond to a certain excited state. Consequently, at our scale, the passage from a state to another is practically instantaneous. No difficulty arises concerning the energy, because on Γ^* (as well as on Γ) the energy is not constant. A very small quantity of energy is sufficient to provoke the passage from Γ on Γ^*. It is unnecessary, as within the conventional language, for ΔE to be instantaneously absorbed.

Relationship between the transition energy and the frequency of the radiation

A radiation is characterized by both its frequency ω and its intensity I. At first sight, we can think that if the radiation causes a change of state of the system by absorption of a certain energy ΔE, this increasing in energy depends on both I and ω. In fact, experiment shows that the intensity is not involved. The frequency only plays a role in the absorption mechanisms, the various frequencies able to be absorbed making a discrete set which characterizes the system under consideration.

Two preliminary points have thus to be find again in our model :

i. the non-intervention of the intensity,

ii. the existence of well-determined frequencies, peculiar to the system.

For the trajectory Γ^*, let us use the same construction procedure as that used for Γ. For an arc C^*, we have the new equation

$$-\tau\dddot{C}^* + \ddot{C}^* - \frac{F(r_0 + C^*)}{m} = \frac{q}{m} E_{rad}^0 \sin(\omega t - \varphi) \tag{X-28}$$

E_{rad}^0 being the amplitude of the electric field and φ the phase of the radiation at the beginning of the burst.

The general solution of this equation is obtained by adding the general solution of the equation corresponding to the absence of radiation (the right-member is equal to zero) to any particular solution of the complete equation.

In the case of the harmonic oscillator of eigenfrequency ω_0, the particular solution is of the form

$$A \sin(\omega t - \varphi) + B \cos(\omega t - \varphi) \tag{X-29}$$

with
$$\begin{cases} A = \dfrac{q}{m} E_{rad}^0 \dfrac{(\omega_0^2 - \omega^2)}{(\omega_0^2 - \omega^2)^2 + \omega^6 \tau^2} \\[4mm] B = -\dfrac{q}{m} E_{rad}^0 \dfrac{\omega^3 \tau}{(\omega_0^2 - \omega^2)^2 + \omega^6 \tau^2} \end{cases} \tag{X-30}$$

The constants A and B are maximum for

$$\omega^2 = \omega_0^2 (1 - \frac{3}{2}\omega_0^2\tau^2) \cong \omega_0^2 \qquad (X\text{-}31)$$

($\omega_0^2\tau^2$ is indeed completely negligible compared to unit) .

In other words, the resonance occurs for the frequency ω_0 of the radiation.

Let us study the sharpness of this resonance. The modulus of the particular solution is

$$\sqrt{A^2 + B^2} = \frac{q}{m} E_{rad}^0 \frac{1}{\sqrt{(\omega_0^2 - \omega^2)^2 + \omega^6\tau^2}} \qquad (X\text{-}32)$$

Thus, at half-maximum, we obtain

$$\omega_{1/2} = \omega_0 \left(1 \pm \frac{\sqrt{3}}{2}\omega_0\tau\right) \qquad (X\text{-}33)$$

i.e. for the total breadth $\Delta\lambda$

$$\frac{\Delta\lambda}{\lambda} = \sqrt{3}\omega_0\tau \qquad (X\text{-}34)$$

For a molecular oscillator, ω_0 is smaller than $10^{15}\,s^{-1}$ and τ is 10^3 times smaller than the characteristic time τ of electron ($6 \times 10^{-24}\,s$) , so that the ratio $\Delta\lambda / \lambda$ is inferior to 10^{-10}. The resonance is therefore extremely sharp.

In the case of any system, the particular solution makes constants A and B appear which will be developed as functions of the powers of E_{rad}^0. The solution will be of the form

$$E_{rad}^0 R_1(\omega) + \left(E_{rad}^0\right)^2 R_2(\omega) + ... = E_{rad}^0 \left[R_1(\omega) + E_{rad}^0 R_2(\omega) + ...\right] \quad (X\text{-}35)$$

We will observe the resonance for

$$\frac{dR_1}{d\omega}(\omega) + E_{rad}^0 \frac{dR_2}{d\omega}(\omega) + ... = 0 \qquad (X\text{-}36)$$

Owing to the ergodicity of the system, the time origin does not intervene, so that the phase φ plays no role. Consequently, on the set of the arcs C^* of Γ^*, the effect will be unchanged if E_{rad}^0 is replaced by $-E_{rad}^0$, which signifies that in the development (X-35) only the even powers of E_{rad}^0 do appear. But $\left(E_{rad}^0\right)^2$ is proportional to the intensity I . Thus the resonance condition (X-35) can be written as follows

$$\mathcal{R}_S(\omega, I) = 0 \qquad\qquad (X-37)$$

$\mathcal{R}_S$ being a function dependent on the system and on the intensity of the radiation (except for the harmonic oscillator) .

An important point has to be notice. The radiation is acting at the level of the elementary arcs C^* and not directly on the complete trajectory Γ^*, which explains that, although the energies on Γ and Γ^* fluctuate, the resonance occurs for a well-determined frequency.

Thus relationship (X-37) defines the frequencies which are able to be absorbed. For small intensities, as it is the case in the practice, this relationship reduces to

$$\mathcal{R}_S(\omega, 0) = 0 \qquad\qquad (X-38)$$

This equation determines the absorption spectrum completely.

The intregration of Eq (VIII-14) which gives the values of the average energies shows that for the harmonic oscillator, ΔE is proportional to ω_0 , i.e., according to what we have seen, to ω. Let us show that such a proportionality to ω is general.

Whatever the origin of the radiation may be, the latter can be considered as being produced by an oscillating electric dipole. At a distance from the source sufficiently great for the radiated wave to be considered as a plane wave, the energy carried by the radiation per time unit is proportional to $e^2\omega^2$, i.e., over a period $\left(\omega^{-1}\right)$, to $e^2\omega$. Now K , thus also $\hbar$, are proportional to e^2. Consequently, over the i -th period after the beginning of the energy absorption, the system absorbs a quantity of energy equal to $b_i\hbar\omega$, b_i being a dimensionless coefficient. After a sufficiently great number of periods for ΔE to be absorbed, we will have

$$\Delta E = \sum_i b_i\hbar\omega = b\hbar\omega \qquad\qquad (X-39)$$

b being a dimensionless coefficient independent of the intensity of the radiation (provided the latter is sufficiently weak) , but which can a priori depend on the nature of the system under consideration. In the case of the harmonic oscillator, $b = 1$, whathever the eigenfrequency of the latter.

Now no dimensionless number can be used to characterize any system. The latter, indeed, is determined, on the one hand, by the mass and the charge of the particles which constitute it, and, on the other, by the potential to which the particles are submitted. This potential is proportional to e^2 / a, a being a length. Thus the ratio potential / square of the charge is the reciprocal of a length. But mass and length are two basic dimensions independent from one another, so that any combination of these properties cannot yield a dimensionless number. Consequently, b is necessarily a universal constant, the same for all the systems. We have seen that it is equal to 1 for the harmonic oscillator. Thus, in all the cases, we will have

$$\Delta E = \hbar \omega \qquad\qquad (X\text{-}40)$$

We will nevertheless notice that this formula refers to an isolated system at rest with respect to the observer. In the practice, the lines of the spectrum are always enlarged by the Doppler effect due to the thermal agitation and by the interaction between the atoms or the molecules [4] .

Molecular spectra

In a molecule the relation (X-40) concerns the whole of the system, i.e. the electrons as well as the nuclei. Let ΔE 's be the corresponding transition energies. The latter form a discrete series.

Given the great difference between the ergodicity times of the electrons and those of nuclei, practically, a series of electronic transitions of energy $\Delta E_e(n)$ corresponds to any position n of the nuclei. Moreover, according to the Born-Oppenheimer approximation (VI-2) , the total wavefunction associated with the molecule can be replaced by the product of a purely electronic function corresponding to the equilibrium position n_0 of the nuclei, and a vibration-rotation function depending on the position of the nuclei, so that the transition energies can be written as follows

$$\Delta E(e,n) = \Delta E_e(n_0) + \Delta E_{nucl.}(n) \tag{X-41}$$

Now the energies $\Delta E(e,n)$ and $\Delta E_e(n_0)$ form discrete series. Therefore the values of the $\Delta E_{nucl.}(n)$ are also discrete. In other words, certain positions n only are acting to give a transition, which significates that the spectrum of the molecule is built up upon families of vibration-rotation lines associated to each electronic transition.

Given that, in general,

$$\Delta E_{nucl.}(n) << \Delta E_e(n_0) \tag{X-42}$$

every thing occurs as if the lines of the electronic transitions were split up into several lines corresponding both to the rotations and the vibrations of the molecule.

Such a structure appears in the diluted gases. In the condensed states, given the increased width of the lines, often no structure is visible. One observes broad bands only.

Utilizable energy carried by a radiation

The relation (X-40) can be interpreted by saying that a radiation of frequency ω carries a *utilizable energy* $\Delta E = \hbar\omega$, independent of its intensity, the word "usable" referring to the possibility for the radiation to modify the state of a system (atom or molecule) by transfer of the energy $\hbar\omega$ when Eq (X-40) is satisfied.

The Einstein idea according to which a particle, the *photon* , must be associated to this energy ΔE is a manner to explain why the intensity of the radiation does not intervene in the quantity of energy which is transferred. The intensity determines the rapidity of the absorption only.

The radiation thus appears as a continuous flow of photons, each of them carrying the same *quantum* of energy $\hbar\omega$. Saying that in certain experiments the photons arrive one by one, only signifies that the intensity of the radiation is extremely small.

Moreover, the existence of this usable energy $\hbar\omega$ explains the photoelectric effect. Indeed, the relation (X-29) determines the value of the threshold from which an electron of the system under consideration becomes free. In any case, it is interesting to remark that in our interpretation, as well as the classical one, the property which determines whether the system will absorb energy or not, is the *quality* of the radiation (namely its frequency) and not its *intensity* .

Consequently, in our model, the photon concept appears as being not necessary. In passing, we will notice that such a conclusion has been obtained by other autors throught different ways [5] .

Induced emission and laser effect

Let us assume that the system is in an excited state corresponding to the trajectory $\Gamma^*(\mathbf{R}+\mathbf{r})$. The radiation induces a modification of the latter. Let $\mathbf{r}'$ be this modification. According to (X-15) we have

$$m\tau\overline{(\ddot{\mathbf{R}}+\ddot{\mathbf{r}}+\ddot{\mathbf{r}}')^2} = q\overline{(\dot{\mathbf{R}}+\dot{\mathbf{r}}+\dot{\mathbf{r}}')(\mathbf{E}+\mathbf{E}_{rad})} \qquad (X\text{-}43)$$

which admits the solution

$$\mathbf{r}' = -\mathbf{r} \qquad (X\text{-}44)$$

Eq. (43) indeed reduces to

$$m\tau\overline{\ddot{\mathbf{R}}^2} = q\overline{\dot{\mathbf{R}}(\mathbf{E}+\mathbf{E}_{rad})} \qquad (X\text{-}45)$$

i.e. according to (X-13), to

$$0 = q\overline{\dot{\mathbf{R}}\mathbf{E}_{rad}} \qquad (X\text{-}46)$$

which is verified given that $\dot{\mathbf{R}}$ and $\mathbf{E}_{rad}$ are non correlated.

In other words, under the effect of the radiation the system returns the energy received during its excitation. This is the *induced* emission which adds to the *spontaneous* emission we have above seen. In passing, we will notice that Eq(44) involves that the transition probabilities are the same for both the absorption and the induced emission.

Moreover, the induced emisssion explains the laser effect. Indeed the latter consists in the desactivation of a metastable state (obtained, for instance, by optical pumpage) with an amplification of the incident radiation. Under the effect of the radiation, the system falls down again on an lower state, emitting an energy equal to that carried by the radiation. This energy adds to that of the incident radiation, from which it results an amplification phenomenon.

Connexion with the perturbation theory

Let us consider a single particle system pertubed by a potential depending on the position of the particle only. Within a reference-frame attached with the particle, the latter is submitted to a perturbation which varies versus time. This perturbation can, at any time, be expanded into a series of monochromatic waves of suitable frequencies which are able to induce jumps from the trajectory of the ground state of the unperturbed system to trajectories corresponding to excited states of the latter.

Then, the state of the system appears as being a mixing of its ground state and its excited states, which justifies the use of a linear combination of the wavefunctions associated to the various states of the unpertubed system to construct the wavefunction corresponding to its new state.

Remark about the states of the continuum

The integration of the Schrödinger equation relative to an atom, for instance, to the hydrogen atom, leads to *closed* states of negative energy, and to a continuum built up upon *not closed* states of positive energy. At first sight, the latter seem as having to be eliminated in the problems concerning the properties of the system in a closed state. Now the perturbation theory shows that it is necessary for them to be introduced in the calculations. The case of the diamagnetism of molecules is typical with this regard. The states of the continuum practically bring half of the contribution in the molecule H_2 when one use the standard atomic orbitals obtained for the isolated hydrogen atom. The situation is physically difficult to be understood within the orthodox formalism where the excited states are defined levels possessing an own existence. Mathematically speaking, the necessity of introducing the continuum arises from that of using a complete basis for the development of the wave-function. All the eigen- functions of the Hamiltonian must be introduced, independently of the fact that they correspond to closed or not closed states.

In our model which is essentially physical, the situation is different. Let us consider the hydrogen atom, for instance. We have seen (Chap. VII) that the motion of the electron can be constructed from arcs of classical trajectories C , the electron jumping from an arc to another. Now these arcs C are of two kinds : The ones are parabolic or hyperbolic of positive energy, i.e. corresponding to not closed states, the other elliptical, of negative energy, corresponding to closed states. Under the effect of the universe field, the electron jumps from an arc C to another whose energy can be either positive or negative. But the electron will describe the reached arc during a very brief time, so that, although the system uses not closed arcs, its average energy will remain negative. Besides, how to explain the value of the average quadratic dispersion we have obtained (Chap. V) for the energy in the hydrogen atom, namely of about $1 a.u.$, whereas the average energy is equal to -0.5 $a.u.$, without introducing positive values of the energy ? Thus we understand why it is necessary to introduce solutions corresponding to the continuum in the orthodox theory where the energy of the various states does not fluctuate.

Thermalization effect

Up to now in this chapter we have only considered systems at the absolute temperature $T = 0$. Let us briefly examine the case where T is different from zero. In this view, we will consider a set of N systems at the temperature T in equilibrium with a thermal radiation corresponding to this temperature. The thermal radiation must be added to the universe field so that, for each system, a new equilibrium state establishes.

According to what we have seen above, the systems pass on excited levels i of average energy $\overline{E_i}$. Owing to the fact that the thermal equilibrium between the systems and the radiation is assumed to be reached, the energy absorbed by a given system is equal to that which the latter returns by radiating, so that the number n_i of systems in the state i is stable.

Consequently, the average energy *per system* is the following

$$\overline{E} = \frac{\sum_{i \geq 0} n_i \overline{E_i}}{\sum_{i \geq 0} n_i} = \overline{E_0} + \sum_{i > 0} \frac{n_i}{N} \Delta E_i \qquad \left(N = \sum_{i \geq 0} n_i \right) \qquad \text{(X-47)}$$

ΔE_i being the energy $(\overline{E_i} - \overline{E_0})$ of the transition $0 \rightarrow i$.

The equilibrium being realized, the ratio n_i / N is equal to the energy density of the radiation $\varepsilon_T(v)$ which depends both on the temperature T and the frequency v, and which is given by the well-known Planck formula [6].

Now this density is vanishing for $T = 0$ whatever the frequency may be. Thus, at $T = 0$, as expected, all the systems are in their ground state. On the contrary, if T is different from zero, the various levels are occupied. On the other hand, the energy density $\varepsilon_T(v)$ tends to zero when the frequency tends to infinity whatever the temperature may be, so that the very high levels are practically not reached by the system. In the greatest number of the cases, the rotation-vibration levels only will be utilized. Electronic transitions will occur only for very high temperatures, e.g. in certain stars.

An important point has to be emphasized : The fact that, at the equilibrium, the populations of the various levels are stable, does not prevent a given system from passing from a state to another, so that if we consider *one* system (and not the set of the systems) , the occupation degrees obtained for the set remain valid on the average over a sufficiently long time for the system under consideration, all the systems exhibiting the same behavior. Here still, the ergodic character of the quantum theory appears when the latter is applied to a unique system. Unfortunately, the direct calculation of the corresponding ergodicity time is difficult to perform. In particular, it depends on the transition probabilities between the various states of the system. Moreover, given that we must take the great number of possible states into account, it is not excluded that in certain cases the ergodicity time reaches macroscopic values.

References

[1] M. Crisp, E. Jaynes, *Phys. Rev.* , **179** (1969) 1253 ; J. Salmon, *Annales Fondation Louis de Broglie* , **15** (1989) 359.

[2] L. Landau, E. Lifchitz, *Mécanique Quantique, Physique Théorique III* (Mir, Moscou) 1966, p. 169 ; A. Julg, *Annales Fondation Louis de Broglie* , **16** (1991) 479.

[3] P. Lochak, C. Meunier, *Multiphase averaging for classical dynamical systems with applications to adiabatic theorems* (Springer, Berlin) 1988.

[4] A. Julg, *J. Chim.Phys.* **53** (1957) 493 ; *Chimie théorique* (Dunod, Paris) 1964 .

[5] W.E. Lamb, M.O. Scully in *Polarisation, Matière et Rayonnement* (Presses Universitaires de France, Paris) 1969; M. Surdin, *Ann. Fond. Louis de Broglie,* **19** (1994) 173.

[6] M. Planck, *Ann. d. Phys.* **4** (1901) 553.

XI

Many-particle systems

Interest of the problem

In order to discuss our interpretation of Quantum Mechanics, up to now we have especially considered systems constituted by one particle or able to be reduced to one particle. A brief incursion into the domain of molecules (i.e. many-electron systems) (Chap. VI) allowed us to obtain interesting results concerning the nature of the chemical bond from the quantum description, but without justifying the latter in the framework of our model. The aim of this chapter is precisely to show how the latter is able to tackle the problem of systems built up upon many particles. The case of many-electron systems within the Born-Oppenheimer approximation (Chap. VI) , of course, will be essentially considered owing to its importance and then we will extend our study to muon-electron systems.

In any case, the study of many-particle systems will constitute a supplementary validity test for our interpretation of Quantum Mechanics in so far as our model will allow not only to find the general results of this theory again, but also to justify the approximation methods used by the latter without other justification than their efficiency.

The Hartree-Fock approximation

Many quantum mechanics textbooks often consider the one-electron systems only, venturing at most to speak of the helium atom. However, quantum mechanics attacked the problem of molecules very early. As far back as 1927, Heitler and London [1] indeed studied the hydrogen molecule. Afterwards, thousands of works have been published in the domain of atoms, molecules and crystals, by using various more or less sophisticated methods.

The most known and the most performing method is the so-called Hartree-Fock method [2] whose logical structure is worth examining.

The Schrödinger equation relating to a many-electron system cannot be directly integrated, from which the necessity of making some simplifying hypotheses resulted. The Hartree-Fock method is based on two hypotheses. First, one puts that in an atom or in a molecule, the electrons occupy well-determined energy levels at most two per level with antiparallel spins. This rule is the so-called Pauli exclusion principle [3] . This point being admitted, one searches for a monodeterminant-like solution (VI-3) built upon one-electron functions which will be determined through a more or less elaborated variation calculation concerning the energy of each electron. This calculation gives a certain number of levels which one fills up according to the Pauli principle from the lowest ones. The most compact filling, i.e. the filling for which the lowest levels are doubly occupied, corresponds to the ground state of the system. Any other filling corresponds to an excited level whose energy is greater than that of the ground state.

Such a method allows to obtain 90-95% of the total electron energy of the system. In order to go beyond this value, one uses the configuration interaction technique. Instead of using a single determinant, one puts that the total wavefunction is a sum of determinants corresponding to the ground state and the excited levels respectively. The gain in energy is called the *correlation energy* . The results are excellent. But the intricacy of the calculations and, above all, their cost entail as consequence that the monodeterminantal approximation is still used in many circumstances.

The success achieved by this standard model based, on the one hand, on the exclusion principle, and, on the other, on the monodeterminantal structure of the wavefunction, asks us the question of the physical meaning of these two approximations. Why would electrons be submitted to a constraint as strong as the Pauli principle? To put this situation as a principle is, obviously, not satisfying. It is true that quantum mechanics does not pretend to justify the axioms on which it is based! On the other, there is the fact that the monodeterminantal approximation is largely sufficient for a very good description of the whole of the electrons to be obtained. Such a structure indeed signifies that a dynamical decoupling between the electrons occurs, each of them moving within the average field created by the other, without being necessary, as it was to be feared, to introduce the positions of the other electrons explicitly. The weak value of the correlation energy compared to the total energy - a few per cent - gives the order of magnitude of the phenomenon.

Given this situation, it is crucial to see whether our model is able to bring any material for an answer to these questions.

Justification of the Hartree-Fock model

First, let us consider the case of a two-electron system. We will notice these electrons 1 and 2 respectively. According to (VII-5) , we can write the corresponding motion equations as follows

$$\begin{cases} -m\tau\dddot{\mathbf{R}}_1 + m\ddot{\mathbf{R}}_1 + \mathbf{F}_{core}(\mathbf{R}_1) + \mathbf{F}_{Cb}(\mathbf{R}_1 - \mathbf{R}_2) = q\mathbf{E} + q\dfrac{\dot{\mathbf{R}}_1}{c} \wedge \mathbf{H} \\ -m\tau\dddot{\mathbf{R}}_2 + m\ddot{\mathbf{R}}_2 + \mathbf{F}_{core}(\mathbf{R}_2) + \mathbf{F}_{Cb}(\mathbf{R}_2 - \mathbf{R}_1) = q\mathbf{E} + q\dfrac{\dot{\mathbf{R}}_2}{c} \wedge \mathbf{H} \end{cases} \quad \text{(XI-1)}$$

where F_{core} is the electron-nuclei electrostatic attraction, and F_{Cb} the Coulomb repulsion force between the two electrons.

Given the ergodic behavior of the system, for each position R_1 of the electron 1, as a first approximation, we can replace the time-dependent Coulomb repulsion $F_{Cb}(R_1 - R_2)$ created by electron 2 on electron 1 by its average value corresponding to the time-independent probability density of electron 2 (assumed to be known) . Under these conditions, we obtain two independent equations

$$\begin{cases} -m\tau\dddot{R}_1 + m\ddot{R}_1 + \Phi(R_1) = qE + q\dfrac{\dot{R}_1}{c} \wedge H \\ -m\tau\dddot{R}_2 + m\ddot{R}_2 + \Phi(R_2) = qE + q\dfrac{\dot{R}_2}{c} \wedge H \end{cases} \qquad \text{(XI-2)}$$

The decoupling between the two variables R_1 and R_2 which results from these equations allows us to assign independent trajectories to the electrons 1 and 2. Let R_1 and R_2 be these trajectories. Now according to what we have seen (Chap. VIII) , the ergodicity involves that for each electron there exist an infinity of equivalent trajectories leading to the same time-average values for *all* the properties associated with the electron under consideration. Consequently, among this double infinity of trajectories, we can arbitrarily select the trajectory R_1 and the trajectory R_2 in order to describe the system.

Let us prove that these trajectories must be different from one another. If the latter would be identical, given the ergodic behavior of the motion of the electrons, we would have

$$R_2(t) = R_1(t + \theta) \qquad \text{(XI-3)}$$

θ being a constant. By substitution in (XI-2) , with respect to time, we would obtain the two following identities

$$\begin{cases} \mathbf{F}(t) \equiv q\mathbf{E}(t) + q\dfrac{\dot{\mathbf{R}}_1(t)}{c} \wedge \mathbf{H}(t) \\[4mm] \mathbf{F}(t+\theta) \equiv q\mathbf{E}(t) + q\dfrac{\dot{\mathbf{R}}_1(t+\theta)}{c} \wedge \mathbf{H}(t) \end{cases} \qquad \text{(XI-4)}$$

$F(t)$ being a suitable function of time, so that θ would be equal to zero. In other words, at any time, the two electrons would be located at the same point, which is physically impossible. Hence

$$R_1 \neq R_2 \qquad\qquad\qquad \text{(XI-5)}$$

Now we have seen that the actual trajectory is the superimposition of a trajectory Γ corresponding to the effect of the electric part of the universe field, and of the Zitterbewegung Z arising from the magnetic part of this field. Symbolically we will put $R = \Gamma Z$. Let us recall that the component Γ is responsible for the dynamical properties and Z for the spin. Owing to the fact that the spins of the electrons 1 and 2 are antiparallel (Chap. IX) , Z_1 is different from Z_2. Consequently, given that R_1 must be different from R_2 (XI-5) , nothing precludes us from choosing $\Gamma_1 = \Gamma_2$. In other words, a possible representation of the system consists in using the same trajectory Γ for each of the electrons i.e. the same space function φ (VI-3) but associated with antiparallel spins, namely $\varphi\alpha$ and $\varphi\beta$.

Now, let us assume that we add a third electron to the system, for instance to a helium atom to form the ion He^- or to an ion Li^+ in order to have the neutral atom Li. The three electrons being physically equivalent to each other, the first idea which comes to mind, is to choose a trajectory Γ_3 associated with the new electron, identical with the first two ones, i.e. that $\Gamma_3 = \Gamma_1 = \Gamma_2$. But $\Gamma_3 Z_3$ must be different from $\Gamma_1 Z_1$ and $\Gamma_2 Z_2$. Given that $Z_3 = Z_1$ or Z_2, necessarily $\Gamma_3 \neq \Gamma_1 = \Gamma_2$.

As a reminder, we will recall that in the Unrestricted Hartree-Fock model [4] the constraint $\Gamma_1 = \Gamma_2$ is not imposed in order to take into account the perturbation caused by the magnetic momentum of the electron

3 on the trajectories ΓZ of the electrons 1 and 2. In this report, we will neglect this correction which is not essential to the discussion.

Whatever that may be,the third electron moving on a trajectory different from those of the first two electrons, a difficulty arises. The three electrons indeed do not exhibit the same average properties. In order to remove this difficulty, it is necessary to admit that the electrons exchange their trajectory. Let us examine more closely how such a process is able to occur.

Let us assume that, at a given time, electrons 1 and 2 describe respective trajectories which, in the absence of any reciprocal interaction, would lead them simultaneously at the same point P. Under the effect of the Coulomb repulsion or more exactly of the difference between this repulsion and the average repulsion used to build the respective trajectories Γ , the two electrons will avoid each other by exchanging their trajectory.

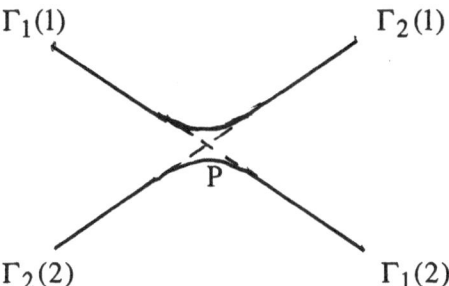

From the arc Γ_1 described by electron 1, the latter passes on the arc Γ_2 and conversely.

Introducing a fourth electron sets no problem owing to the fact that the new trajectory can accept two electrons with antiparallel spins. If the system possesses more than four electrons, it will be necessary to introduce new trajectories.

Exchanging two electrons is very rare at the scale of the particles, but it is very frequent at our scale, so that, over a time interval which will seem to us as being extremely brief, the electrons will exhibit the same average properties. In other words, the trajectory followed by a given electron is the *union* of the various trajectories Γ_i. This trajectory is therefore the same for all the electrons of the system under consideration.

A point has, nevertheless, to be made precise concerning the spin of the electrons during the passage from a trajectory to another. Let us assume that, before exchange, electron 1 follows trajectory Z_1 (Chap. IX) associated with the trajectory Γ_1, and that electron 2 follows trajectory Z_2 (different from Z_1) associated with Γ_1, electron 3 trajectory Z_1 associated with Γ_2 and electron 4 trajectory Z_2 associated with Γ_2. We will symbolize this situation by $(\Gamma_1 Z_1, \Gamma_1 Z_2, \Gamma_2 Z_1, \Gamma_2 Z_2)$.

Let us assume that electrons 1 and 3 exchange their trajectory. Then, we will have the following situation $(\Gamma_2 Z_1, \Gamma_1 Z_2, \Gamma_1 Z_1, \Gamma_2 Z_2)$ which is compatible with the Pauli principle. On the contrary, if electrons 1 and 4 exchanged they trajectory, we would obtain $(\Gamma_2 Z_1, \Gamma_1 Z_2, \Gamma_2 Z_1, \Gamma_1 Z_2)$, which is forbidden by the exclusion principle. Consequently, the exchange between two electrons is possible only if their spins are parallel to each other. In other words, the spin momentum is kept, which may be logically foreseen.

This extremely important restriction allows us to understand the Hartree-Fock result concerning the energy. Let us consider a system described by the determinant $\left| ...\varphi_i^2...\varphi_j^2... \right|$, i.e. corresponding to an even number of electrons using the orbitals φ_i by pairs with antiparallel spins. The energy associated to an electron using the orbital φ_i is the following

$$e_i = I_i + J_{ii} + 2\sum_{j\neq i}\left(J_{ij} - \frac{1}{2}K_{ij}\right) \qquad \text{(XI-6)}$$

I_i being the sum of the kinetic energy of the electron using this orbital and of the interaction energy between the electron under consideration and the nuclei; J_{ii}, the average Coulomb repulsion between the two electrons using the orbital φ_i

$$J_{ii} = \int \varphi_i^2(1)dv_1\int\frac{1}{r_{12}}\varphi_i^2(2)dv_2 \qquad \text{(XI-7)}$$

J_{ij}, the average repulsion between two electrons using the orbitals φ_i and φ_j respectively

$$J_{ij} = \int \varphi_i^2(1)dv_1\int\frac{1}{r_{12}}\varphi_j^2(2)dv_2 \qquad \text{(XI-8)}$$

and K_{ij} the so-called exchange integral

$$K_{ij} = \int \varphi_i(1)\varphi_j(1)dv_1\int\frac{1}{r_{12}}\varphi_i(2)\varphi_j(2)dv_2 \qquad \text{(XI-9)}$$

For a two electrons system, ($\left|\varphi_i^2\right|$), Eq(XI-4) reduces to

$$e_i = I_i + J_{ii} \qquad \text{(XI-10)}$$

which corresponds to the energy of an electron within the average field created by the nuclei and the other electron.

If the system contains a more great number of electron pairs, the first idea is that the total energy is t

$$e_i = I_i + J_{ii} + 2\sum_{j\neq i} J_{ij} \qquad \text{(XI-11)}$$

As a matter of fact, we have to subtract a so-called exchange term. Its origin is easy to be understood in our model. Indeed, in the vicinity of the point P where exchange occurs between two trajectories, the kinetic energy increases. Indeed, let us consider the idealized case of the elastic diffusion of two electrons which would reach the point P at the same time with the same velocity V. Let Δ_1 and Δ_2 be their respective trajectories. In the absence of interaction between these electrons, the latter would be diffused with the velocity equal to V along the directions Δ'_1 and Δ'_2 which coincide with the ones of Δ_2 and Δ_1 respectively. Now let us assume that for each electron the effect of the Coulomb repulsion can be compared to the one of a repulsive impulsion which would be acting only at the very time of the impact. By symmetry, the corresponding vectors $m\mathbf{v}$ will be oriented along the exterior bisectrix of the angle (Δ_1, Δ_2) with opposite directions

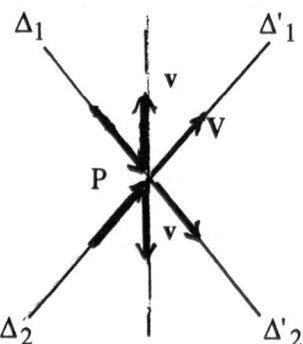

The resultant velocity ($\mathbf{V} + \mathbf{v}$) is in modulus greater than the initial one $\mathbf{V}$. It follows an increasing in kinetic energy.

According to the virial theorem applied to a coulombian system $(2\overline{T} + \overline{U} = 0$, i.e. $\overline{E} = -\overline{T})$, the average energy decreases. On the other hand, we have seen that exchange is possible only if the spins of the two electrons remain parallel. Consequently, one of two orbitals will be concerned, from which the factor 1/2 which appears in the exchange term K_{ij} corresponding to the gain in stability arising from the non-crossing. In the Hartree-Fock calculation, this factor 1/2 is the direct consequence of the monodeterminantal structure adopted for the wave-function which our model thus justifies *a posteriori* .

Connection between the spin and the Fermi-Dirac statistics

It is well-known that electrons are fermions, i.e. obey the Fermi-Dirac statistics which governs indistinguishable particles constrained to occupy various states at the rate of *one* by state only. On the other hand, according to Quantum Mechanics, electrons exhibit a spin I=1/2, which involves that the component of the latter along a given direction is equal to $\pm \hbar / 2$.

The Relativistic Quantum Mechanics shows that the choice of the statistics is directly connected with the value of I [5], in particular that all the particles of spin equal to 1/2 are fermions. The proof of this result is not elementary given the complexity of the theory. On the contrary, in our model, such a connection can be easily established.

Indeed, on the one hand, we have seen that, at any time, each electron of any atom or of any molecule uses a trajectory (Γ Z) different from those followed by the other electrons. Moreover, the electrons of the system are unceasingly exchanging between one another so that over a sufficiently long time they are indistinguishable. Thus we are in the applicability conditions of the Fermi-Dirac statistics (indistinguishability of the particles and monoparticle fulling of the states Γ Z).

On the other, in an atomic or molecular system the spins of the electrons using a given trajectory Γ are opposite, so that every thing occurs as if the spins can be oriented along two opposite directions only.

These two properties, namely the choice of the statistics and the two orientations of the spin are, in our model, the consequences of the existence of the universe field and not of any intrinsic property of the particles. In other words, the connection which exists between the spin and the statistics is indirect contrary to what the Relativistic Quantum Mechanics states.

Coming back on the orbital domains

In order to interpret the chemical bond (Chap. VI) , we have introduced orbital domains [6] in which the electrons reside by pairs during a certain time before exchanging one another. As we have said, such domains have not a clear physical interpretation since they overlap on another in the space $\{r\}$. The present discussion allows us to make the physical meaning of these domains precise. They are identifiable indeed with the various trajectories Γ_i described by the electrons. It is, nevertheless, to be noticed that the stay time of an electron inside a given domain cannot be identified with the corresponding ergodicity time. The example of the atomic orbitals we are going to see, clearly shows the difference between these two times.

Slater's rules

Slater [7] showed that within the central field approximation, an atom can be described by a determinant or a sum of determinants built from functions whose angular portions are identical with those of the exact solutions obtained for the hydrogen atom, but with radial parts corresponding to apparent nuclear charges $Z_{app.}$ smaller than the actual charge Z of the nucleus. The identity of the angular portions with those of the hydrogen atom orbitals allows us to keep the notation : $1s$, $2s$, $2p$ ...

First, let us class the orbitals into successive *groups* :

$(1s)(2s,2p)(3s,3p)(3d)(4s,4p)...$

In order to obtain the apparent nuclear charge acting on an electron of a given orbital of this group, one subtracts the following quantities from the actual nuclear charge Z :

0.35 for each of the other electrons belonging to this group

(except for $1s$ where one subtracts 0.31),

0.85 for each electron of the inner groups if the orbital under consideration is of type s or p ; 1 if it is of type d or f,

0 for the electrons of the outer groups.

Such a rule which is at variance with the fact that all the electrons of the system are equivalent, can be justified in our model as follows. The electrons are localized within orbital domains which practically correspond to the various atomic orbitals (See the discussion about the beryllium atom in Chap. VI) , forming spherical concentric shells around the nucleus. The average potential acting on an electron during its stay within a given domain, is that created by the nucleus, minus the potential created by the inner electrons and those of the same group, whereas the electrons of the outer groups have no action according to the well-known Gauss theorem in electrostatic.

The identification of the so-defined atomic orbitals with the various orbital domains allows us to make the difference between the stay time and the ergodicity time precise. For these domains, indeed, the average quadratic velocity varies as Z^2 / n^2 (n = principal quantum number of the hydrogenlike orbital to which the orbital under consideration corresponds) and the characteristic size of the domain behaves as n^2 / Z . Consequently, the ergodicity time varies as $n^3 / Z_{app.}^2$. The ergodicity time of the various domains are different from one another whereas the stay times are the same because when an electron goes out of a domain, another electron replaces it immediately.

Hund's rule

Let us consider the case where - according to the Hartree-Fock terminology - the two less bound electrons can use not *one* as above but *two* levels a and b of the same energy. We will assume that the other electrons are associated by pairs with antiparallel spins, so that they can be ignored in the discussion.

Two cases are possible :

i. The electrons use two different levels with parallel spins, from which the two following configurations

$$(\Gamma_a Z_1, \Gamma_b Z_1) \text{ and } (\Gamma_a Z_2, \Gamma_b Z_2)$$

ii. The electrons use the same level with opposite spins, the corresponding configurations being

$$(\Gamma_a Z_1, \Gamma_a Z_2) \text{ and } (\Gamma_b Z_1, \Gamma_b Z_2)$$

If there was no exchange between the trajectories Γ_a and Γ_b, these four configurations would possess the same energy. But in the first case, for each configuration exchange between the trajectories Γ_a and Γ_b is possible, hence a decreasing in energy, the same for the configurations. Owing to the fact that the spins are opposite, no exchange is possible between these configurations, thus no interaction can occur so that the obtained states will coincide with the respective corresponding configurations.

On the contrary, in the second case, exchange between the two configurations is possible. There is thus an interaction between the latter with a symmetrical splitting for the energy, according to the well-known rule in mechanics of coupled systems of the same nature. Now the energy which is involved in this exchange is the same as in the first case. It arises indeed in the two cases from the passage from trajectory Γ_a to the trajectory Γ_b or conversely. The decreasing in energy is therefore the same so that the state of lower energy is threefold degenerate. This is precisely what the Hund rule says, namely that the electrons use the maximum number of levels of the same energy with the maximum spin multiplicity. In the present case, we have a *triplet*.

The preceding proof can be easily generalized to the case of two levels which exhibit energies different from one another, and which are simply occupied as this occurs for an one-electron excitation. We obtain a singlet and a triplet, the latter possessing the lower energy.

Muon-electron systems

Muon is a particle of the same charge as electron but with a much greater mass (ca. 200 times) . As we have seen in Chapter VI, Quantum Mechanics applies to this particle. In particular, we can expect the existence of muonic systems, i.e. atoms or molecules whose electrons would be replaced by muons. For instance, from hydrogen atom $H(p^+,e^-)$, p^+ being the hydrogen nucleus, we will obtain $H(p^+,\mu^-)$. Such systems have effectively been prepared in accelerators [8] . All the results obtained for the normal hydrogen atom are formally valid for this new atom under the condition to replace the electron mass m_e by the one m_μ of muon. Thus, if, as a first approach, we consider that the nucleus p^+ remains at rest, we obtain the following muonic Bohr radius

$$a_\mu = \frac{\hbar^2}{m_\mu e^2} = \frac{m_e}{m_\mu} a_e \cong \frac{1}{200} a_e \qquad (XI-12)$$

Likewise, for the energy

$$E_\mu = -\frac{e^2}{2a_\mu} = \frac{m_\mu}{m_e} E_e \cong 200 E_e \cong -100 a.u. \qquad (XI-13)$$

In other words, the μ-density is strongly concentrated in the vicinity of the nucleus and the system is much more stable than that with an electron. The taking into account of the finite mass of the nucleus does not qualitatively modify these results.

The problem becomes more complex if we consider atoms containing both electrons and muons, for instance the ion $H^-(p^+,e^-\mu^-)$ which corresponds to $H^-(p^+,2e^-)$. The question is to know how these particles behave in such a system.

By symmetry the densities corresponding respectively to each of these particles are spherical, as well as the average potential which is acting on each of them. Consequently, according to what we have seen for the usual atoms, as a first approximation, each of these particles can be described by a $1s$ hydrogenic orbital with a suitable effective nuclear charge $Z_{eff.}$. If the particles were independent of one another, the effective nuclear charges would be all the same and equal to the charge of the nucleus, namely +1 . Given the large difference between the two Bohr radii, the muonic density and the electron density do not overlap, that corresponding to the muon being very concentrated in the near vicinity of the nucleus, that corrsponding to the electron forming a shell much more distant from the nucleus. Now we have seen that for a $1s$ orbital the ergodicity time (for a given Z) is proportional to the corresponding Bohr radius (V-13) . Given the values of these radii, the ergodicity time of the muon is much smaller than the one of the electron so that, practically, at any time, the field which is acting on the electron can be replaced by that created by an electric point charge, very close to -1 , located at the nucleus. It follows that the effective nuclear charge of the orbital associated to the electron is very small, which makes the ergodicity time of the electron still greater, increasing the difference between the two ergodicity times, and consequently reinforcing the validity of the approximation made in replacing the muon by a charge -1 located at the nucleus.

Consequently, the muon and the electron of the system will be described by the product of two functions $1s$ whose effective charges are $Z_{eff.}(\mu^-) = 1$ and $Z_{eff.}(e^-) \cong 0$ respectively. No exchange between the particles has to be introduced because the latter are of different nature. Given that the effective nuclear charge acting on the electron is very close to zero, the electron will be very weakly linked, so that spontaneously the ion $H^-(p^+, e^- \mu^-)$ will disintegrate and the reaction

$$H(p^+, e^-) + \mu^- \longrightarrow H(p^+, \mu^-) + e^-$$

will be irreversible.

In atom $He(\alpha, e^-\mu^-)$, where α is the helium nucleus He^{2+}, $Z_{eff.}(\mu^-) = 2$ and $Z_{eff.}(e^-) \cong 1$, the substitution of an electron by a muon makes the system more stable

$$He(\alpha, 2e^-) + \mu^- \longrightarrow He(\alpha, e^-\mu^-) + e^-$$

Thus the situation is quite comparable with that which occurs in molecules where the Born-Oppenheimer approximation (VI-2) allows to split up the total wave-function into two parts corresponding to the nuclei and the electrons respectively.

References

[1] W. Heitler, F. London, Z. *Physik* **4** (1927) 455.

[2] D.R. Hartree, *Proc. Cambridge* **24** (1928) 89, 111, 426 ; V. Fock, *Zeit. Physik* **61** (1930) 126 ; J.C. Slater, *Quantum Theory of atomic Structure* (McGraw-Hill, New York) 1960.

[3] W. Pauli, *Zeit. Phys.* **31** (1925) 765.

[4] For instance, see : R. McWeeny, B.T. Sutcliffe, *Methods of Molecular Quantum Mechanics* in *Theoretical Chemistry* , vol. 2 , edited by D.P. Craig and R. McWeeny (Academic Press, London & New York) 1969 , and references therein.

[5] L. Landau, E. Lifchitz, *Physique Théorique, Théorie quantique relativiste I* , (Mir, Moscou) 1973, p. 113.

[6] A. Julg, P. Julg, *Int. J. Quantum Chem.* **13** (1978) 483.

[7] J.C. Slater, *Phys. Rev.* **36** (1930) 57.

[8] For an extensive monography, see : C.S. Wu, W. Hughes, *Muon Physics* (Academic Press, London & New York) 1977.

XII

The wave-particle duality

Origin and interpretation of the concept

Within the classical framework, the notion of particle as well as that of wave do not set problem. An electron is a material object just like a pebble. A wave is essentially the propagation of a deformation. A physical entity cannot be both wave and matter.

The question of the undulatory or corpuscular nature of a physical phenomenon seems to be appeared for the first time with regard to light. For Newton, the latter was constituted by small corpuscles whereas for Huyghens, it was a wave. The two points of view seemed to be defensible at this time. Afterwards, the second conception prevailed over the first one in the nineteenth century following the Fresnel works, before being questioned by Einstein at the beginning of the twentieth century about the interpretation of the photoelectric effect.

On the contrary, the situation seemed to be clear for electrons. They are particles ... up to the day when Louis de Broglie [1] foresaw their diffraction by a crystal, just as for the X-rays which were considered as being undulatory in nature. One knows that experiment confirmed this prediction [2]. The same phenomenon was subsequently observed with protons and neutrons. Any particle, therefore, exhibited the dual nature of both a wave and a particle.

In fact, the wave-particle duality foreseen by Louis de Broglie arose from the formal analogy between the equation of the motion of a particle and that of the propagation of a wave. From this analogy it follows that the wave of wavelength

$$\lambda = \frac{h}{p} \qquad \qquad \text{(XII-1)}$$

has to be associated to any particle of linear momentum p . This fundamental wavelength is called the de Broglie wavelength. It is involved in the calculations of the diffraction spectra.

Now, subsequently, quantum mechanics introduced the Schrödinger function Ψ. The question was then to know the connection between the de Broglie wave and the Schrödinger function.

For Schrödinger, the wave Ψ had a real character, the particle being not localized. On the contrary, Born denied any physical character to this wave, considering that it was only a probability representation attached to the particle. For de Broglie, it was necessary to consider two completely distinct solutions : the one, u , of physical nature, not normalizable and exhibiting a singularity corresponding to the particle, the other, Ψ - i.e. the Schrödinger solution - normalizable and without singularity, and which would be a probability representation only. In fact, the interpretation which triumphed was the abstract one of Bohr (1927) , namely that of the *complementarity* of the two aspects, assigning to the Schrödinger function Ψ the role of a probability representation which allows to predict the result of the measurement, abandoning any clear idea concerning the wave as well as the particle, the latter being both relegated to the position of out-of-date concepts.

Towards 1951, Louis de Broglie, nevertheless, took his idea of the double solution again, trying to make it precise from an idea expressed by Bohm (1956) [3]. Roughly, his conception is the following : The particle follows well-defined trajectories inside its wave, jumping from one trajectory to another under the effect of the random interaction with a hidden *subquantal* medium which would play the role of an invisible thermostat. For Louis de Broglie, an indeterministic element seemed to be necessary for finding the meaning of Schrödinger's wave again, u being proportional to Ψ out of the domain in which the particle is localized. In others words, the Louis de Broglie model attempted to restore the physical meaning both of the particle and of the wave, contrary to the Bohr opinion.

In our model, the problem is set in a completely different manner given that it is based, on the one hand, on the inalienable physical nature of electrons as well as of protons, and, on the other, on the essentially undulatory nature of the electromagnetic waves, in particular of light. Now we have seen that such hypotheses do not preclude to obtain a quantum-like formalism so that it is not necessary to dress up the electron in an undulatory character, the obtained function Ψ (VIII-14) resulting in fact from the purely mathematical correspondence between the physical dynamics of the particle and the operator algebra defined by $\overline{G} = \langle \Psi \hat{G} \Psi \rangle$

(VIII-1). In this outlook, this function would have no physical character.

However, it is important to emphasize that, finally, our conception is very close to that of de Broglie. In the latter indeed the (rather mysterious) subquantal medium induces random jumps from a possible trajectory to another. In our model, the situation is quite analogous because a random element is introduced, namely the action of the universe field which explains the jump of the particle from an arc C to another. But, in our model, this random effect is physically clear in origin, given that the universe field arises from the whole of the systems which constitute the universe. Conceptually, our model is thereby more satisfactory than the de Broglie interpretation. The wave-particle duality, in fact, is only the result of an unjustified mixture of a physical phenomenon, namely the actual motion of the particle with its mathematical transcription.

In any case, whatever the interpretation may be, it is necessary to explain the physical origin of the diffraction of the particles (electrons, neutrons or protons) and that of complex systems (helium atoms [4] or hydrogen molecules [5]) by slits or crystals, even if these particles or these systems pass one by one through the diffracting device, drawing the same diffraction pattern on a distant screen as that which would be obtained by a wave whose wavelength is given by the de Broglie relationship (XII-1) [6] .

Before discussing this problem, we will in first examine the case of a free particle.

The spreading of a wave-packet

The problem of a free particle moving along a right-line from a given point with the initial velocity v_0 is a classical exercise in the conventional quantum mechanics. In this view, one considers a wave-packet of a very limited extend at the time $t = 0$, represented for convenience by an extremely sharp Gauss-function corresponding to the superimposition of monochromatic waves which, by interference, give a probability practically equal to zero at any point of the straight-line, except around the point where the particle is located at the time $t = 0$.

By using the time-dependent Schrödinger equation, one shows, on the one hand, that the center of gravity of the wave-packet is moving with the velocity v_0 - which is consistent with the inertia principle - and, on the other, that the wave-packet is spreading while keeping its gaussian form so that, after a sufficiently long time, the probability for finding the particle at a given point becomes the same at any point of the straight line, and thereby tends to zero. The particle is then completely delocalized. More precisely, if at $t = t_0$, the breadth at half-maximum of the distribution is equal to σ_0, at the time t the latter will be equal to

$$\sigma = \frac{\hbar t}{2m\sigma_0} \tag{XII-2}$$

which involves that $\sigma_0^2 = \hbar t_0 / 2m$ (m = mass of the particle).

In other words

$$\frac{\sigma}{t} = \frac{\sigma_0}{t_0} \qquad\qquad\qquad (XII\text{-}3)$$

The breadth of the distribution varies thus as t .

This result sets a problem of interpretation. Two explanations are possible :

i. The classical notion of particle is not valid, the wave-function only has a physical character. Under these conditions, unacceptable values for the size of the particles are obtained : e.g., for an electron coming from Sun, given that the travel takes about 3 days ($3 \times 10^5 s$) , we obtain a number of kilometers for σ practically equal to the number $1/\sigma_0$, σ_0 being expressed in centimeters ! Now, errors and ignorance excepted, such a size is not observed for these electrons.

ii. On adopts the statistical interpretation, and one considers that we deal with a great number of electrons starting at the same time with the same velocity, and that during the travel the packet of particles spreads out.

In our model the interpretation is the following. If the universe field did not exist, the motion of the particle would be uniform : $v = v_0$. Owing to this field, fluctuations superimpose to this motion. From the initial situation ($x_0 = 0$, $\dot{x}_0 = v_0$), under the effect of the successive impulses of averagelife time τ ,we have the following velocities:

$$v_0 , \; v_0 + \delta\dot{\xi}_1 , \;\; v_0 + \delta\dot{\xi}_1 + \delta\dot{\xi}_2 , \text{ etc.}$$

The displacements ξ corresponding to each impulse being proportional to characteristic time τ associated with the particle (VII-20) , after the nth impulse the position of the particle is

$$\chi_n = n\tau v_0 + \tau\Big[(n-1)\delta\dot{\xi}_1 + (n-2)\delta\dot{\xi}_2 + \ldots + \delta\dot{\xi}_n\Big] \qquad (XII\text{-}4)$$

Hence, within the reference-frame moving with the velocity $v_0 << c$, a displacement x_n given by

$$x_n = \chi_n - n\tau v_0 = \tau\Big[(n-1)\delta\dot{\xi}_1 + \ldots\Big]$$

$$\cong (n-1)\delta\xi_1 + (n-2)\delta\xi_2 + ... + \delta\xi_n \qquad \text{(XII-5)}$$

Thus, for the square of this relative displacement, if we retain the terms which appear in the calculation of the average value only, we obtain

$$x_n^2 \cong (n-1)^2 \delta\xi_1^2 + (n-2)^2 \delta\xi_2^2 + ... \cong n^3 \overline{\delta\xi^2} \qquad \text{(XII-6)}$$

Now, according to (VII-17), $\overline{\delta\xi^2} \cong \left(\dfrac{q\tau_0}{mc}\right)^2 \overline{A^2}$.

Hence, the average value of x^2 after the nth impulse

$$\overline{x^2} \cong n^2 \overline{\delta\xi^2} \cong \left(\frac{t}{\tau_0}\right)^2 \frac{q^2\tau_0^2}{m^2c^2} \overline{A^2}$$

$$\cong \left(\frac{t}{\tau_0}\right)^2 \frac{q^2}{mc^3} \frac{c\tau_0^2}{m} \overline{A^2} \cong \frac{t^2\tau K}{m\tau_0^2} \qquad \text{(XII-7)}$$

which signifies that, starting from an infinitely sharp distribution, after a sufficiently long time t , the distribution exhibits a finite extend characterized by

$$\sigma \cong \frac{t}{\tau_0}\sqrt{\frac{\tau K}{m}} \qquad \text{(XII-8)}$$

with $K = \hbar / 2$ (VIII-21). For instance, for an electron $\sigma(cm) \cong 10^{-12} t(s)$, which shows that the perturbation arising from the universe field is completely negligible concerning the motion of an electron in our laboratory experiments. In other words, in such experiments, electron behaves as a point particle. Given that σ varies as $m^{-1/2}$ the effect is still weaker for a proton and *a fortiori* for a macroscopic body, which justifies the applicability of classical mechanics at our scale.

Let us assume that at time t_0, sufficiently great for the asymptotical law (XII-8) to be applicable, the breadth at half-maximum is equal to σ_0. At the time t , the latter will be

$$\sigma(t) = \frac{\hbar t}{2m\sigma_0}\left(\frac{\tau t_0}{\tau_0^2}\right) \qquad \text{(XII-9)}$$

Apart from the factor $\left(\tau t_0 / \tau_0^2\right)$, this relationship is the same as that obtained by the quantum theory. Now the value of this factor is independent of the adopted time unit. We will choose $\tau_0 = \tau$. If we refer to the breadth at the time $t_0 = \tau_0 = \tau$, the two formulae, then, become identical with each other.

A remark, nevertheless, has to be made. In our model, if all the particles started at the same time with the same velocity, it is obvious that no dispersion would occur. Consequently, in order to account for the observed extend of the distribution, it is necessary for the particles not to start exactly at the same time. Such a condition is *ipso facto* realized for particles coming from solar explosions or from those of supernovae. In spite of the brevity of these explosions (a few milliseconds) , the interval between the start of two successive particles is extremely great with respect to the characteristic time τ.

Wave associated with a particle

From the foregoing, it follows that the motion of the particle along the x -axis is never rigorously uniform. Fluctuations superimpose to the uniform motion which would correspond to the average velocity of the particle. Let $V \ll c$ be this velocity. Within a reference frame moving with this velocity, for each impulse received from the universe field, the fluctuation is such as relationship (VII-18) is verified, so that, according to what we have seen in Chapter VIII, at any time, the fluctuation can be assimilated with harmonic oscillations.

Let v be the frequency to which, at a given time, the fluctuation corresponds. The total energy carried by the electron is equal to the sum of its kinetic translation energy $(1/2)mV^2$ and of its oscillation energy $h\,v\,/2$.

Let us assume that, at this time, the particle is stopped by a receptor. This total energy will use by the latter to induce the passage to an excited state of energy $h v_0$ (with respect to its initial state) , v_0 being an absorption frequency of the receptor. Then

$$\frac{1}{2}mV^2 + \frac{1}{2}h\nu = h\nu_0 \qquad \text{(XII-10)}$$

(The transverse fluctuations have not to be taken into consideration. Indeed, they exert no pressure on a screen perpendicular to the x -direction)

Now the frequency ν of the electron will be absorbed only if it is equal to ν_0 , so that, from (XII-10)

$$\nu_0 = \frac{mV^2}{h} \qquad \text{(XII-11)}$$

A frequency ν' different from ν_0 has no effect on the receptor. In other words, there exists a privileged frequency which, energetically speaking, can be attached with the electron in its translation motion at the velocity V . By introducing the wavelength λ_0 corresponding to the propagation of the oscillations at the velocity V , we obtain

$$\lambda_0 = V/\nu_0 = h/mV \qquad \text{(XII-12)}$$

which is exactly the de Broglie wavelength.

Nevertheless, we will notice the purely formal association of a wave with the particle which, contrary to the de Broglie ideas, does not involve the double nature (undulatory and corpuscular) of the particle.

Electron diffraction

In order to simplify the problem, let us consider the case of electrons passing through two parallel slits cut in a wall. According to the orthodox interpretation, an electron coming in the vicinity of the slits loses its corpuscular nature and changes into a wave which can simultaneously pass through the two slits without difficulty. The wave diffracts from the two slits giving the well-known diffraction pattern on the screen.

In our model in which electrons remain true particles, the interpretation is the following. In the vicinity of the slits, the universe field is not isotropic, it changes so as to reflect the presence of the slits. More precisely, according to the classical theory of diffraction, given that the universe field can be decomposed into Fourier integrals, we obtain the superimposition of stationary interference systems, each of them being characterized by its wavelength λ.

Now the interaction between the electron and the universe field will be maximum if in its translation-oscillation motion the electron remains in phase with one of the components of the field, which involves that only the component corresponding to the wavelength λ_0 will be acting on the electron. Moreover, given that the interaction is all the stonger as the amplitude of the field is great, the probability to find the electron at a given point of a screen located at a sufficiently large distance of the slits, is proportional to the intensity predicted at this point by the diffraction of a radiation of wavelength λ_0 (XII-12).

In fact, an infinity of trajectories are able to be used by the electron. The choice of the trajectory is determined by the small random deviations of the velocity with respect to the x - axis arising from the components of the universe field perpendicular to this axis. It follows that after a great number of impacts on the screen, we obtain the same pattern as that which would correspond to the de Broglie wave.

Consequently, contrary to the conventional interpretation, electrons are finally the *detectors* of a pre-existent local deformation of the universe field only. The mysterious wave-particle duality is thus completely removed.

No problem, of course, arises concerning the light which is in our model undulatory in nature, the concept of photon being abandoned. The incident wave has a sufficiently large extend for passing through the two slits simultaneously.

The particle in a box

The problem of a particle in a box is a generalization of the one of the diffraction by two slits we have above considered. Owing to the presence of the walls of the box, inside the latter the universe field forms an interference system. Given the energy exchanges between the field and the particle, the latter reaches an equilibrium state of positive energy.

For simplicity, we will consider a particle of mass m inside a rectangular box with edges of length a, b, c (with $0 < x < a$; $0 < y < b$; $0 < z < c$). Let $A_x(t,x)$ be the component of the vector potential along the direction x. The equation which governs the motion of the particle along this direction is

$$-\tau m\ddot{x} + m\dot{x} = qA_x(t,x) \tag{XII-13}$$

Now a is a scale factor for the interference system. The amplitude of the field, indeed, varies inversely as a. Consequently the energy of the particle (which is reduced to its kinetic energy) behaves as $\dfrac{1}{2m}(m\dot{x})^2$ i.e. as $\dfrac{1}{ma^2}$. On the other hand, if the universe field would be equal to zero, $\hbar$ would be equal to zero and the particle at rest. Therefore, the energy of the particle is proportional to a power of $\hbar$. By homogeneity, this energy is proportional to $\hbar^2$. Consequently, the energy of the particle in the box is the following

$$E_1 = \frac{B\hbar^2}{m}\left(\frac{1}{a^2} + \frac{1}{b^2} + \frac{1}{c^2}\right) \tag{XII-14}$$

We find the quantum result by putting $B = \pi^2/2$.

Now, let us consider a box of the same size but divided into two compartiments by a removable wall located at $x = \lambda a$ ($\lambda < 1$) and let us assume that the particle is enclosed inside the compartiment ($\lambda a \times b \times c$). According to (XII-12), its energy is

$$E_\lambda = \frac{B\hbar^2}{m}\left(\frac{1}{\lambda^2 a^2} + \frac{1}{b^2} + \frac{1}{c^2}\right) > E_1 \tag{XII-15}$$

Let us remove the dividing wall. From this time, the particle can visit the whole box $(a \times b \times c)$. According to Quantum Mechanics, the particle passes on a *non-stationary* state described by a linear combination of the time-dependent eigenfunctions corresponding to the cavity $(a \times b \times c)$, its energy remaining the same, i.e. equal to E_λ.

In our model, the situation is different. Indeed, owing to the energy exchanges between the field and the particle, a new equilibrium state of energy E_1 (XII-14) is reached. In other words, the removing of the dividing wall (which does not require energy) entails a decrease of the energy of the particle equal to $(E_\lambda - E_1)$, the energy lost by the particle being recovered by the universe field.

Momentum associated with an electromagnetic radiation. Application to the Compton effect

Let us consider a system immersed within an electromagnetic radiation. The latter exerts a force on it. According to classical electrodynamics [7], the corresponding momentum density is equal to $(\mathbf{E} \wedge \mathbf{H}) / 4\pi c$ and the energy-flux density to $(\mathbf{E} \wedge \mathbf{H})c / 4\pi$ so that, if the system absorbs the energy W, the total mechanical momentum is equal to W /c. This relationship is valid whatever the actual absorption process and the time necessary for the radiation to be absorbed. In particular, it is applicable to the utilizable energy $h\nu$ carried by an electromagnetic radiation of frequency ν (Cf. Chap. X)

$$p = \frac{h\nu}{c} \tag{XII-16}$$

If the system is reduced to a free electron, by writing the conservation of energy and momentum, we obtain the well-known equations which allow to account for the Compton effect. The latter is thus explained without being necessary for the photon concept to be introduced.

Closed and unclosed systems

In all the preceding chapters, we have assumed that the systems remain closed, i.e. that the electrons describe trajectories such that, after a sufficiently long time, it is possible to speak of an equilibrium state, the average values of the various properties of the system becoming stable. In this case, quantum mechanics applies to *one* particle. On the contrary, in the problems treated in this chapter, namely the case of the free particle and the diffraction or interference phenomena, each of the particles works on its own, the result predictet by quantum mechanics corresponds to a set of a great number of particles submitted to random actions. In other words, the *unclosed* systems are relevant to a *statistical* or rather *collective* interpretation. Our model thus brings a clear answer about the statistical or individual nature of quantum mechanics by distinguishing the case of closed systems from that of unclosed systems.

References

[1] L. de Broglie, Thesis, Paris, 1924 ; *Annales de Physique,* **3** (1925) 22.

[2] G. Davidson, L.H. Germer, *Phys. Rev.* **30** (1927) 705 ; G.P. Thomson, *Proc. Roy. Soc.* **A 117** (1928) 600.

[3] L. de Broglie, in L. de Broglie, J.L. Andrade de Silva, *La réinterprétation de la mécanique ondulatoire 1* (Gauthier-Villars, Paris) 1971.

[4] O. Carnal, J.M. Mlynek, *Phys. Rev. Lett.* **66** (1991) 2689.

[5] I. Esterman, O. Stern, *Z. Phys.* **61** (1930) 95.

[6] J. Faget, *Revue d'Optique* **40** (1961) 347.

[7] J.D. Jackson, *Classical Electrodynamics* (John Wiley, New York) 2th ed., 1975, p.238.

XIII

Microreversibility and ergodicity

The specific character of the time variable

Time is certainly one of the notions which is the most difficult to be apprehended. This arises essentially from the fact that it is impossible to realize a material standard as for masses and lengths. Time is perceived as "some thing" which is flying without going back.

Many authors have discussed the problem of time, in particular, in the old days, Aristotle (ca. 350 B.C.) who said that time is the measure of the motion, and St Augustinus (ca. 400 A.C.) for whom time can be conceivable only by referring to the universe which is continually in evolution. In our age, the philosophical content of the notion of time has been studied by Bergson with respect to that of duration in a famous but questioned survey of the Einstein theory of relativity [1] .

In order to make this essentially intuitive notion of time precise and to get free it from its anthropomorphic character, Mechanics considers time as a parameter which enters in the motion equations. An important point, nevertheless, has to be emphasized. Contrary to the biological time which we feel and which is elapsing out without a possible return, the time introduced in Mechanics can vary from $-\infty$ to $+\infty$, without constraint concerning the direction of its variation.

Reversibility and irreversibility in Mechanics

First, let us consider the simple case of a harmonic oscillator. Let x_0 and v_0 be the position and the velocity of the particle respectively, at the time $t = 0$. The trajectory C followed by the particle is defined by the following equations

$$\begin{cases} x(t) = x_0 \cos \omega t + \dfrac{v_0}{\omega} \sin \omega t \\ v(t) = v_0 \cos \omega t - \omega x_0 \sin \omega t \end{cases} \qquad \text{(XIII-1)}$$

ω being the oscillator frequency.

At the time $t = -\theta$, the state of the system was defined by

$$\begin{cases} x(-\theta) = x_0 \cos \omega \theta - \dfrac{v_0}{\omega} \sin \omega \theta \\ v(-\theta) = v_0 \cos \omega \theta + \omega x_0 \sin \omega \theta \end{cases} \qquad \text{(XIII-2)}$$

Let us assume that at the time $t = 0$, we reverse the velocity of the particle : $v_0 \rightarrow -v_0$. Then the trajectory $\tilde{C}$ followed by the particle is defined by the following equations

$$\begin{cases} \tilde{x}(t) = x_0 \cos \omega t - \dfrac{v_0}{\omega} \sin \omega t \\ \tilde{v}(t) = -v_0 \cos \omega t - \omega x_0 \sin \omega t \end{cases} \qquad \text{(XIII-3)}$$

At the time $t = \theta$, we will have

$$\begin{cases} \tilde{x}(\theta) = x_0 \cos \omega \theta - \dfrac{v_0}{\omega} \sin \omega \theta \\ \tilde{v}(\theta) = -v_0 \cos \omega \theta - \omega x_0 \sin \omega \theta \end{cases} \qquad \text{(XIII-4)}$$

i.e.

$$\tilde{x}(\theta) = x(-\theta) \text{ and } \tilde{v}(\theta) = -v(-\theta) \qquad \text{(XIII-5)}$$

whatever θ may be.

In other words, the trajectory $\tilde{C}$ is identical with C, but from the point x_0 it is described in the opposite direction. We will say that the motion is *reversible* . We will nevertheless note that if the particle follows the trajectory C in the opposite direction, in no case we will invert the course of time. The situation is identical with that which occurs when one rewinds a film with the same speed as for the projection.

Now, let us consider a damped oscillator

$$\ddot{x} + 2f\dot{x} + \omega^2 x = 0 \tag{XIII-6}$$

whose solution is

$$x = \left[x_0 \cos \omega t + \frac{v_0 + fx_0}{\omega} \sin \omega t \right] e^{-ft} + O(f^2) \tag{XIII-7}$$

$O(f^2)$ being terms of equal or superior to 2 order, negligible if f is very small with respect to ω.

If we reverse the velocity at the point x_0, the new trajectory $\tilde{C}$ is defined by

$$\tilde{x}(t) = \left[x_0 \cos \omega t + \frac{-v_0 + fx_0}{\omega} \sin \omega t \right] e^{-ft} + O(f^2) \tag{XIII-8}$$

From (XIII-7) and (XIII-8) we have

$$\tilde{x}(\theta) \neq x(-\theta) \quad \text{and} \quad \tilde{v}(\theta) \neq -v(-\theta) \tag{XIII-9}$$

The particle does not pass again at the same points with opposite velocities. This result was physically foreseeable. Indeed, owing to the friction term, the oscillator does damp down. The evolution of the system is *irreversible* . Whatever the initial conditions may be, the oscillator tends towards a limiting state, the immobility in the present case. Thus, the friction force appears as being the deep cause of this ineluctable march of the system to a well-determined state. The impression we have, namely that time is elapsing without possible return, results from this.

Algebraically, relation (XIII-5) which can be considered as being the reversibility criterium, arises from the invariance of the motion equation when one replaces t by $-t$:

$$Eq(-t) = Eq(t) \tag{XIII-10}$$

Friction, irreversibility and stability

Let us now consider a damped oscillator submitted to an external harmonic force of frequency α (cf VIII-26)

$$\ddot{x} + 2f\dot{x} + \omega^2 x = a \sin \alpha t \tag{XIII-11}$$

Owing to the friction force, the reversibility condition (XIII-5) is not verified. Consequently, the trajectories $\tilde{C}$ are not identical with the trajectories C. As in the previous case (XIII-6) , the system tends to a limiting state in which the time average values of the various dynamical properties are constant. This state, nevertheless, does not correspond to the immobility as in the preceding case for which we have $x = A\sin\alpha t + B\cos\alpha t$. The properties of this state are determined, on the one hand, by the value of f and, on the other, by those of a and α . In particular, the amplitude x will be maximum for a certain frequency very close to ω. This is the resonance phenomenon. If the friction coefficient f was equal to zero, on the resonance, the amplitude x would be infinite. The system would disintegrate.

More generally, if the system is submitted to a set of external forces whose frequencies form a quasi-continuous series, the system will evolve to a state which is the superimposition of the various states corresponding to the various frequencies α . This state is stable and the amplitude x remains finite. On the contrary, if $f = 0$, for the resonance frequency, the system disintegrates.

This elementary example clearly shows the connection which exists between the friction forces and the stability of the system. Now we have seen that irreversibility and stability (which involves an ergodic behavior) are two strongly connected phenomena which arise from the existence of friction. Moreover, in passing, we will notice that in the limiting state reached by the system, the energy arising from the external medium is compensated by that dissipated by the friction.

Similarity to our model

The case of the damped oscillator submitted to external harmonic forces has to be compared with the model we have developed in Chapter

VII. Indeed owing to the electromagnetic damping force condition (XIII-10) is not verified, hence the irreversible behavior of the system. On the other hand, the latter is unceasingly excited by the variations of the universe field . Under the conjugated effect of these two factors, whatever the detailled evolution of the system, the latter tends ineluctably to a state characterized by the stability of the time average values of the various dynamical properties. In particular, the radiated energy is, on the average, compensated by that received from the rest of the universe. This limiting state can be considered as a *strange attractor* which drives the system whatever its initial conditions may be. In this connection, let us recall that as far back as 1971, Louis de Broglie considered the stationary states of Quantum Mechanics as being privileged states in which the system remains preferentially. See also Ref [2] .

But the point the most important to be remarked is that the universe field arises itself from the damping electromagnetic forces, so that, finally, the irreversibility of the system and its stability (and consequently its ergodic behavior) are the consequence of a unique phenomenon, namely the Lorentz damping force. Without these forces, the systems would disintegrate. Our conclusion is more general than the one of Fer [3] for whom the irreversibility is the foundation of the stability of the physical world. The irreversibility does not appear to us as being the first cause of the behavior of matter because it arises from the friction forces.

Parallelism with entropy

The evolution of the macroscopic systems is governed by *entropy* . Spontaneously, any system evolves so that its entropy increases. It passes through more and more probable and less and less well-ordered states up to a final state characterized by a maximum disorder. The case of a crystal we vaporize is typical in this regard. In the crystal each atom possesses a well-defined position whereas in the gas, all the atoms are geometrically and dynamically equivalent over a sufficiently long time, i.e. possess the same average position and the same properties.

In our model, starting from an arbitrary situation, the system tends more or less quickly to a state for which, on the average, the dynamical properties are constant. This is true for systems constituted by one particle (formally at least) whose average position becomes stable, as well as for many-particle systems whose constituting particles become geometrically and dynamically equivalent. As an example of single particle system we can quote the one of an optically active molecule (Chap. IV) , and for a many-particle system, the one of the electrons in a molecule which are completely delocalized on the whole system.

Consequently, it is allowed to ask whether, starting from a certain state at $t = 0$, we can find a function which would indicate the progress of this irreversible march towards the ergodic equilibrium state. In Quantum Mechanics the knowledge of the density is sufficient for obtaining the values of all the dynamical properties of the system. Consequently, we can think that the function we search, is a function of the difference between the quantum density which corresponds to the ergodicity, and the density corresponding to the state of the system at the time under consideration. We will put

$$\Omega(t) = \int \left[\frac{\rho_Q(M) - \overline{\rho(M)}_t}{\rho_Q(M)} \right]^2 \rho_Q(M) dM \qquad \text{(XIII-12)}$$

$\overline{\rho(M)}_t$ being the average density corresponding to the state under consideration at the point M over the time interval $(0, t) : \frac{1}{t} \int_0^t \rho(M)_\theta \, d\theta$;

$\rho_Q(M)$ is the quantum density at the point M . The integrating is extended over the whole space.

On the other hand, we will remark that function Ω (XIII-12) can be written as follows

$$\Omega = \int \left[1 - \frac{\overline{\rho(M)_t}}{\rho_Q(M)} \right]^2 \rho_Q(M) dM$$

(XIII-13)

$$= \int \rho_Q(M) dM - 2\int \overline{\rho(M)_t} dM + \int \frac{\overline{\rho(M)_t}^2}{\rho_Q(M)} dM$$

i.e. owing to the fact that the densities are normalized to unit

$$\Omega(t) = \int \frac{\overline{\rho(M)_t}^2}{\rho_Q(M)} dM - 1$$

(XIII-14)

The ergodicity is reached when $\overline{\rho(M)}$ is equal to the quantum density $\rho_Q(M)$ at any point of space. At this time, Ω is equal to zero. In all the other cases Ω is strictly positive.

Let us assume that we start from a state in which the particle is at rest at the point M_0. The density $\rho(M)$ is equal to zero at any point of space, except at the point M_0 where it can be represented by a Dirac distribution, so that we can write

$$\Omega = \frac{1}{\rho_Q(M_0)} \int \delta^2 (M - M_0) dM - 1$$

(XIII-15)

which is infinite.

Likewise, if we consider a particle in a box with a density $\overline{\rho(M)_t}$ constant at any point of the box (as in a perfect gas) , Ω is infinite given that the quantum density ρ_Q is equal to zero on the walls of the box. Consequently, the function Ω can vary from $+\infty$ to zero.

In order to illustrate the definition of Ω , we will consider a pair of enantiomers A and B. To simplify the calculations, we will use the double-well potential proposed in Eq(IV-6) . Let $\{A\}$ and $\{B\}$ be the wells corresponding to the active forms A and B respectively.

Starting from the configuration A, let us assume that at the time t the system is still localized in the well $\{A\}$. In other words, the form A can be isolated (cf. arsines, Chap. IV) . The corresponding density over the time interval $(0, t)$ is the one which corresponds to a harmonic oscillator centred on the well $\{A\}$.

Then

$$\overline{\rho(M)} \cong \begin{cases} 2\rho_Q & M \in \{A\} \\ 0 & M \in \{B\} \end{cases} \tag{XIII-16}$$

$$\Omega_A \cong \int_{\{A\}} \frac{4\rho_Q^2}{\rho_Q} dM -1 = 2 \int_{\{A+B\}} \rho_Q dM -1 = 1 \tag{XIII-17}$$

The system which is localized inside the well $\{A\}$ has not reached the ergodicity : Ω is different from zero. As the system jumps from one well to the other, the time average density tends progressively to the quantum one, and Ω to zero.

Another interesting example is given by the behavior of an excited state (in the meaning of Quantum Mechanics), assumed as being stable. In this view, we will consider a particle constrained to move on the segment $(0,a) : 0 \le x \le a$.

In the ground state

$$\rho_Q(x) = \psi_0{}^2(x) = \frac{2}{a} \left(\sin \frac{\pi x}{a} \right)^2 \tag{XIII-18}$$

In the first excited state

$$\overline{\rho_M(x)} = \psi_1{}^2(x) = \frac{2}{a} \left(\sin \frac{2\pi x}{a} \right)^2 \tag{XIII-19}$$

Therefore

$$\Omega_1 = \frac{2}{a} \int_0^a \frac{(\sin 2\pi / a)^4}{(\sin \pi x / a)^2} dx -1 = 1 \tag{XIII-20}$$

This result has to be compared with the conclusion obtained in Chapter X, namely that an excited state is unstable. Spontaneously, the latter does evolve towards the ground state for which Ω is equal to zero.

Consequently, the so-defined function Ω appears as being able to give information about the gap between the state under consideration and the limiting one of ergodic behavior which corresponds to the quantum description. Starting from $+\infty$ for a localized particle, Ω is decreasing up to zero. A priori, of course, nothing allows us to assert that this decreasing is monotonic versus time. This function is decreasing in the mean only.

Moreover, we will remark that $(-\ln\Omega)$ varies from $-\infty$ to $+\infty$. Consequently, this function behaves in the mean as the thermodynamical entropy which is increasing when the system spontaneously evolves. Therefore, we can consider this function as a *micro-entropy* , the prefix *micro* reminding that this function is defined for a single system whereas in Statistical Mechanics the entropy S applies to a set of systems. This difference explains the monotonic behavior of the entropy. Indeed, the fact that the latter applies to a great number of systems, S is automatically averaged over the whole of the systems so that its variation seems to be monotonic whereas, at the scale of the systems, S is not necessarily increasing. Its variation exhibits fluctuations.

Whatever that may be, the preceding remarks clearly show the fundamental role played by the electromagnetic damping forces in the irreversibility of the phenomena at the microscopic scale, and in the irreversible march of time which arises from it. More generally, it is allowed to think that the macroscopic irreversibility also arises from friction forces, the word "friction" being taken *largo sensu* , i.e. from any interaction which dissipates energy. Now all the macroscopic interactions can be theoretically reduced to elementary electromagnetic interactions so that the irreversibility of the macroscopic phenomena we observe and the irreversible march of time would finally arise from the electromagnetic damping forces, and not from any specific property of time.

Reference

[1] H. Bergson, *Durée et simultanéité* (1922) reedited by Presses Universitaires de France (Paris) 1968 . For a recent discussion, see M. Cariou, *L'atomisme, Bergson et Lucrèce* (Aubier, Paris) 1978.

[2] J.-P. Eckmann, D. Ruelle, *Rev. of Modern Physics* **57** (1985) 617.

[3] F. Fer, *L'irréversibilité, fondement de la stabilité du monde physique* (Gauthier-Villars, Paris) 1977.

XIV

Does Planck's constant vary versus time ?

The problem of the past variability of the fundamental parameters in physics

In our model, the constant K (VII-20) , we have identified, apart from a numerical factor, with the Planck constant (VIII-21) , is appeared as being strongly connected with the generalized interaction between all the systems which constitute the universe. If, as the standard big-bang model claims, the universe is effectively expanding, it is logical to wonder whether $\hbar$ is an absolute constant or is varying versus time.

In fact, the problem of the past variability of the Planck constant cannot be separated from that of the other fundamental constants, namely the charge e and the mass m of electron, the mass M of proton, the light velocity c , the gravitation constant G , the constants of the strong and weak nuclear interactions (g_s and g_w resp.) , the Boltzmann constant k , and so on. The Plank constant, indeed, does appear ever associated with one or several of these constants.

The problem of the past variability of these parameters has been tackled by various authors $[1-11]$. Nevertheless, in general, not all of the parameters were considered as being able to vary. In the present discussion, which has been in part published in Ref $[12]$, we will admit that *all* the parameters are able to vary.

A preliminary remark

At first, we will remark that no direct experiment is able to conclude about a possible variation of quantities whose dimension is different from zero (e.g.: e , m , h). Indeed, let us consider a certain property U . Let u be the corresponding unit. The value indicated by the measurement device is a dimensionless number :

$$[U] = \frac{U}{u} \qquad \text{(XIV-1)}$$

Hence, even if the observed value $[U]$ exhibits a variation versus time, it is impossible to distinguish the part which is due to the variation of U itself from that arising from the variation of u , and, conversely, the observed value can be constant, even if U varies.

In order to make the relationship between these quantities precise, we will consider two successive measurements performed at times θ_0 and $\theta_0 + \Delta\theta$ respectively. Our device will indicate the following variation of U

$$[\Delta U] = [U_0 + \Delta U] - [U_0] \qquad \text{(XIV-2)}$$

U_0 being the value of U at θ_0 .

At $\theta_0 + \Delta\theta$, the value of the unit u is different from u_0 its value at θ_0 . It is equal to $(u_0 + \dot{u}_0\Delta\theta)$, $\dot{u}_0$ being the derivative of u at θ_0 so that we obtain the apparent variation

$$[\Delta U]_{app} = \frac{U_0 + \Delta U}{u_0 + \dot{u}_0\Delta\theta} - \frac{U_0}{u_0} = \frac{\Delta U}{u_0} - \frac{\dot{u}_0}{u_0^2} U_0\Delta\theta \qquad \text{(XIV-3)}$$

On the other hand, the apparent time-interval between the two measurements is

$$[\Delta t]_{app} = \frac{\Delta\theta}{\tau_0 + \dot{\tau}_0\Delta\theta} = \frac{\Delta\theta}{\tau_0} \qquad \text{(XIV-4)}$$

τ_0 being the value of the time unit at θ_0 , and $\dot{\tau}_0$ its derivative, so that one has

$$\left[\frac{\Delta U}{\Delta t}\right]_{obs} = \left[\frac{\Delta U}{u_0} - \frac{\dot{u}_0}{u_0^2} U_0\Delta\theta\right] \frac{\tau_0}{\Delta\theta}$$

i.e., when $\Delta\theta$ tends to zero

$$\left(\frac{dU}{dt}\right)_{obs}^{0} = \left(\frac{dU}{dt}\right)_{act}^{0} - \left(\frac{\dot{u}}{u}\right)_{act}^{0} U_{act}^{0} \qquad \text{(XIV-5)}$$

the indices *obs* and *act* refer to the *observed* value and to the *actual* one respectively.

For the relative variation rate $(\dot{U}/U)$ [12], we obtain

$$\left(\frac{\dot{U}}{U}\right)_{obs}^{0} = \left(\frac{\dot{U}}{U}\right)_{act}^{0} - \left(\frac{\dot{u}}{u}\right)_{act}^{0} \qquad \text{(XIV-6)}$$

This relationship shows, in particular, that for a dimensionless property, the observed variation rate coincides with the actual one $(\dot{u}/u=0)$. Consequently, only dimensionless numbers are able to give direct information concerning the behavior of the fundamental parameters versus time.

Experimental data

A certain number of universal invariants is reveled by experiment :

i. The study of the relative doublet splitting of optical emission lines in radiogalaxies leads to the following variation rate of the fine structure constant $\alpha = \dfrac{e^2}{\varepsilon_0 \hbar c}$: $|\dot{\alpha}/\alpha| < 10^{-15}$ year^{-1} [8].

ii. The comparison between the optical redshifts of hydrogen atoms and those corresponding to heavier atoms in old radiogalaxies indicates that the ratio m/M had between 0.7 and 1.4 times its present value [9], which corresponds to a relative variation rate of about $\pm 10^{-11}$ yr^{-1}. In fact, the average value of this ratio is practically equal to 1.0, so that the variation rate is certainly weaker in absolute value.

iii. The study of desintegration products in the natural fission reactor of Oklo (Gabon) [8] shows that for the ratio $\beta = g_w M^2 c/\hbar^3$: $|\dot{\beta}/\beta| < 10^{-12}$ yr^{-1}.

On the other hand, some experiments are put forward as a proof of the invariability of h and k . In fact, their interpretation is mistaken. For instance, the fact that the energy of photons of a given color does not depend on their age [13] cannot be interpreted as showing that h is constant. It shows only that the law $E = h \nu$ is remained valid since the early universe [11] . Likewise for k : The fact that the fossil radiation which fills the universe corresponds very exactly to that of a black body at 2.9 K , only proves that the ratio h /ak (a = length unit) is remained constant. No conclusion can be drawn concerning k itself contrary to what certain statements [14] .

Choice of a unit system

First of all, we must make the adopted unit system precise. In this discussion, we will use the following system built up on the electron characteristics

i. *Length unit* : We will adopt the Bohr radius corresponding to an hydrogen atom whose nucleus would be infinitely heavy

$$a = \frac{\varepsilon_0 \hbar^2}{me^2} \tag{XIV-7}$$

This choice is equivalent to that of a certain emission wavelength of a given atom (e.g. : Cs) . Indeed

$$\frac{e^2}{\varepsilon_0 a} \propto h\nu = \frac{hc}{\lambda} \text{ , i.e. } a \propto \lambda \tag{XIV-8}$$

owing to the fact that α is constant.

i.i. *Mass unit* : The electron mass, m .

i.i.i. *Time unit* : We will choose

$$\tau = \frac{e^2}{\varepsilon_0 mc^3} \tag{XIV-9}$$

i.e., according to (XIII-8) : $\tau \propto a / c$ $\hspace{2cm}$ (XIV-10)

Instead of defining this unit from electron characteristics, we can also define it from gravitation phenomena by putting

$$\tau_g \propto \sqrt{\frac{a^3}{Gm}} \tag{XIV-11}$$

In order to preserve the structure of the time-space interval in relativity $ds^2 = \Omega^2(t)\left[c^2 dt^2 - dx^2 - dy^2 - dz^2\right]$ it is necessary that the two definitions (XIV-10) and (XIV-11) coincide with one another, i.e. that

$$Gm \propto ac^2 \tag{XIV-12}$$

Now according to a theorem due to Cartan [15] , the ratio G/c^2 is time invariant, so that (XIV-12) simplifies as follows

$$m \propto a \tag{XIV-13}$$

i.v. *Charge unit* : The absolute value of the electron charge, $|e|$.

Derived units

From these basic units, we can define derived units. For instance:

Force: $mc^2/a \propto e^2/\varepsilon_0 a^2$ $\varepsilon_0 : e^2/amc^2$

Speed: $a/\tau \propto c$ $\mu_0 = 1/(\varepsilon_0 c^2) : am/e^2$

Acceleration : $c/\tau \propto a/\tau^2$ Magnetic charge: $amc/|e|$

Action : $mac \propto ma^2/\tau$ Electric field: $mc^2/a|e|$

Energy : $mc^2 \propto e^2/\varepsilon_0 a$ Magnetic field: $|e|c/a^2$

$$\tag{XIV-14}$$

Invariance of the physical laws

Physical laws have been established at a certain time. But they must remain true at all the epochs, otherwise they would lose any meaning. It is easy to verify that this invariance is preserved within the chosen unit system. As an example, let us consider the fundamental dynamical relation $F = M\gamma$. From (XIV-1) and (XIV-14) ,we get

$$F = [F]mc^2/a \ ; \ M = [M]m \ ; \ \gamma = [\gamma]c^2/a$$

Hence $[F]=[M].[\gamma]$. Consequently, at any time, even if the properties themselves and the units vary, we will always obtain the same relationship between the observed values $[F]$, $[M]$ and $[\gamma]$. Here is the very sense of the permanence of the physical laws.

This invariance can be verified for all the laws. We have seen the case of $E = h$ ν . It arises from the fact that our unit system is coherent. But, in any case, this invariance does not involve that of the parameters themselves.

This invariance can be applied to the Boltzmann distribution. In first, we will define a temperature scale T from the phase transition of a pure substance, e.g. by adopting $T_0= 273.16$ degrees for the triple point of water. To this end, we will write that the cohesion energy of the crystal $\left(\propto e^2 /\varepsilon_0 a\right)$ is equal to kT_0 . Thus k is proportional to $e^2 / \varepsilon_0 a$, i.e.

$$hc \propto ak \qquad\qquad\qquad (XIV\text{-}15)$$

Moreover, it is easy to verify that

$$hc \propto a^4\sigma \qquad\qquad\qquad (XIV\text{-}16)$$

σ being the Stephan constant.

Invariance of the light velocity and that of G

We have seen that the speed unit is proportional to c , so that, if we apply relationship (XIV-5) to the speed v of a particle, we obtain

$$\gamma_{obs} = \gamma_{act} -\left(\frac{\dot{c}}{c}\right)v_{act} \qquad\qquad (XIV\text{-}17)$$

γ being the acceleration of the particle.

Let us assume that, within a certain reference frame, the particle is at rest. Within another frame moving with the speed V with respect to the first frame, the motion of the particle is uniform ($v = V$). Now, according (XIV-17), the motion would be uniformly accelerated. In order to remove the inconsistency, it is necessary that the light velocity is constant $\left(\dfrac{\dot{c}}{c} = 0\right)$.

Consequently, according to Cartan's theorem, G is constant.

First consequences

It results from the invariance of α and from relationships (XIV-7) , (XIV-13) , (XIV-15) and (XIV-17) that

$$\hbar \propto \frac{e^2}{\varepsilon_0} \propto k^2 \propto a^2 \propto m^2 \propto \sigma^{-1} \tag{XIV-18}$$

G and c being constant [12] .

Connection with the strong and weak interactions

The time invariance of the ration m/M allows to establish a connection between the electromagnetic interaction and the strong and weak nuclear interactions. The strong interaction, indeed, is characterized by the constants g_s , such as

$$g_s^2 \propto a M_s c^2 \tag{XIV-19}$$

M_s arising from the interaction. Now the proton mass arises from the electromagnetic interaction between the quarks which constitute it, and from the strong interaction between these particles. The invariance of the ratio m/M and that of β involves

$$g_s^2 \propto \hbar \text{ and } g_w \propto \hbar^2 \tag{XIV-20}$$

We will remark that these relationships entail the proportionality between the electron mass and that of any nucleus, and, consequently, of any macroscopic body. This point is very important, because it allows to identify the unit system we have chosen with the conventional systems, for instance with the International System.

On the analogy of the electromagnetic fine structure constant α , the theory of the interactions introduces the following fine structure constants

$$\alpha_s = g_s{}^2 / \hbar c \; ; \; \alpha_w = g_w m^2 c / \hbar^3 \text{ and } \alpha_G = G M^2 / \hbar c \quad \text{(XIV-21)}$$

associated with the strong and weak interactions, and with gravitation respectively. According to (XIV-18) and (XIV-20), the ratios are time independent as α itself.

Lastly, the masses of the bosons $W_\pm$ and Z_0, associated with the weak interaction, and which are proportional to $\dfrac{\hbar}{c}\sqrt{\dfrac{e^2}{g_w}}$ [16] , behave as the electron mass.

Consequently, we can make the proportionality relationships (XIV-18) complete as follows [12]

$$\begin{cases} \hbar \propto \dfrac{e^2}{\varepsilon_0} \propto k^2 \propto a^2 \propto m^2 \propto \sigma^{-1} \propto g_s{}^2 \propto g_w{}^{1/2} \\ m \propto M \propto M_W \propto M_Z \end{cases} \quad \text{(XIV-22)}$$

c and G remaining constant.

In passing, we will remark that the ratio electromagnetic interaction / gravitation interaction (e^2 / ε_0) / $Gm\,M$ (i.e. the first of Dirac Large Numbers [4]) is time-invariant, as this author assumed.

The existence of these relations between properties which are very different from one another, can appear, at first sight, as being rather surprising. In fact, they are the consequence (if not the proof !) of the connection which exists between the electromagnetic, strong and weak interactions, and gravitation .

Origin of the time-invariability of α

According to relationships (VII-19) and (VIII-21) , $\hbar$ is proportional to $c\tau_0{}^2[\mathbf{A}-\mathbf{A_0}]^2$. Now, given the form of the vector potential we have adopted, a and e are scale factors for the variations of the field of the universe. Consequently, $\overline{[A - A_0]}^2$ is proportional to $\left(\dfrac{e^2}{a^2}c\tau_0\right)^2$, so that

$$\hbar \propto c\tau_0{}^2 \overline{[A - A_0]}^2 \propto \frac{e^2}{c}\left(\frac{c\tau_0}{a}\right)^4 \propto \frac{e^2}{c} \tag{XIV-23}$$

which shows that the ratio $\alpha = e^2 / \hbar c$ or rather $e^2 / \varepsilon_0 \hbar c$ (by re-introducing the coefficient ε_0) is time-independent.

This result is very important, because it constitutes a supplementary argument in favor of our model.

The principle of conservation of energy

According to what precedes, the past variability of the fundamental parameters of physics cannot be a priori excluded, provided that the relationships (XIV-22) are respected. Such a variability, nevertheless, can appear as being incompatible with the principle of conservation of energy which is considered as being one of the most solid pillars of physics. For instance, the energies of atoms which are proportional to $e^2 / \varepsilon_0 a$ i.e. behave as a , would not be kept if the constants varied.

In fact, such a variability is possible in our model because no system is strictly isolated so that its energy can vary. The *total energy* of the universe only must be kept.

Anyhow, experiment allows us to reach the number (XIV-1) which measures the energy in the units of the epoch, and not the energy itself. Given that the various energies are of the form $e^2 / \varepsilon_0 a$ or mc^2 which both behave as m , and that the energy unit is mc^2 (XIV-14) , according to (XIV-6) the variation rate *observed* for the energy is equal to zero, so that the *apparent* energy is kept, although, in an *absolute* manner, the latter varies.

Variation of $\hbar$ versus the expansion of the universe

The universe is not unchanging. After a brief but tumultuous phase in which particles and the first nucleons appeared, the universe entered a quieter period characterized by a regular expansion which, for a sufficiently old matter-dominated universe, is governed by the following equation [17]

$$2\frac{\ddot{R}}{R} + \frac{\dot{R}^2}{R^2} + \frac{Kc^2}{R^2} = \Lambda \qquad (\text{XIV-24})$$

where Λ is the cosmological constant; K a parameter equal to +1, -1 or 0, according as the universe is closed with a positive curvature, open with a negative curvature, or flat (open and euclidian). R is the curvature radius of the universe (or the scale factor if $K = 0$).

The behavior of the universe depends strongly on the values of Λ and K. If Λ is equal to zero, the universe is ever-expanding if it is open, but expanding and recontracting alternatively if it is closed. On the contrary, if Λ is different from zero, in the two cases (open or closed universe) , it is expanding forever.

At the present time, the value of Λ is extremely weak, practically equal to zero, whereas, according to recent works, during the primeval phase of the universe, Λ exhibited an enormous value, which would explain the big-bang phenomenon [18] .

Now, in the absence of matter which would curve the space, the latter must be euclidian ($K = 0$) without past or future, which involves that Λ would be equal to zero. Consequently, we are entitled to think that Λ tends to zero as the matter dilutes under the effect of the expansion of the universe.

On the other hand, it seems that a closed universe ($K = +1$) is abler to account for a certain number of properties than an open universe, in particular the average universe isotropy by eluding the difficulty arising from the synchronism of the phenomena at two points of space arbitrarily removed from one another. Therefore we will adopt this model.

If $K = +1$ and Λ is constant and nonzero, for a sufficiently old universe, R is increasing as an exponential function of time [17] .

On the contrary, if Λ tends to zero versus time, owing to the fact that Kc^2/R^2 tends to zero, R asymptotically tends to the solution of the Einstein-de Sitter model [17] which precisely corresponds to K and Λ equal to zero

$$R(t) \propto t^{2/3} \tag{XIV-25}$$

Therefore we will put

$$R(t) = At^{2/3} + a\varphi(t) \tag{XIV-26}$$

$\varphi(t)$ being a function of t tending to zero when t tends to infinity, and A and a coefficients.

Let us assume that $\varphi(t)$ behaves as t^{-n} when t tends to infinity. By substituting in (XIV-24) we obtain

$$\Lambda = \frac{c^2}{R^2} + \frac{2a(9n^2 + 3n - 2)}{9t^{(3n+8)/3}} + \ldots = \frac{c^2}{R^2} + \frac{2a(9n^2 + 3n - 2)}{9(R/A)^{4+3n/2}} + \ldots \tag{XIV-27}$$

If $n = \dfrac{2}{3}$, we obtain

$$\Lambda = \frac{c^2}{R^2} + \frac{8a}{9}\left(\frac{A}{R}\right)^5 + \ldots \quad \text{and} \quad R = At^{2/3}\left(1 + \frac{a}{At^{4/3}} + \ldots\right) \tag{XIV-28}$$

and if $n = \dfrac{4}{3}$

$$\Lambda = \frac{c^2}{R^2} + 4a\left(\frac{A}{R}\right)^6 + \ldots \quad \text{and} \quad R = At^{2/3}\left(1 + \frac{a}{At^2} + \ldots\right) \tag{XIV-29}$$

By aesthetics the value $n = 4/3$ seems to be better than $n = 2/3$. In both the cases, Λ behaves as R^{-2}i.e. as $t^{-4/3}$.

It is not obvious, of course, that this solution is the true one. Nevertheless this calculation shows that a closed universe can be in eternal expansion even if it closed, contrary to what occurs if Λ remains constant.

For such a universe, the mass and the energy are finite. Now the total energy of the universe depends, on the one hand, on the fundamental parameters and on the number of particles of each species at the time under consideration, and, on the other, on its curvature radius R at this time. Moreover, according to the phenomenological relationships (XIV-22) we have established, the fundamental parameters can be expressed as a function of one of them, $\hbar$ for instance. On the other hand, along the path followed by the universe, the number of particles of each species is a function of time, i.e., finally, of R . Consequently, the total energy of the universe can be expressed as a function of $\hbar$ and R only. From which it results

$$\Phi(\hbar, R) = Cst \qquad\qquad\qquad\qquad (\text{XIV-30})$$

because the total energy of the universe must remain constant.

This relationship shows that if $\hbar$ and the other parameters which are connected to it, remained constant, R would be well determined, i.e. that the universe would be static. Consequently, the "constants" do vary versus time.

Now, physically, the charges and the masses cannot become infinite so that, when R tends to infinity, $\hbar$ and the other parameters tend to finite limits which a priori can be different from zero or equal to zero.

On the other hand, according to (XIV-1) and (XIV-14) , the energy of the universe is equal to the value $[E_{univ.}]$ we would observe (if the measurement was possible !) multiplied by the energy unit mc^2. The total energy of the universe remaining constant, successive measurements would always give the same value $[E_{univ.}]$ whereas the energy unit varies versus time. Consequently, the total energy would vary, which is inconsistent with the conservation of energy, *unless* $[E_{univ.}]$ be equal to zero. Thus, Eq (XIV-24) reduces to $\Phi(\hbar, R) = 0$.

Now when R tends to infinity, the interaction between the various systems of the universe become negligible compared with the internal energies of the particles which are proportional to mc^2. Likewise, the radiation energy, proportional to $\sigma T^4 R^3$, behaves as a^2 / R, thus tends to zero owing to the fact that the temperature T of the universe is proportional to a / R (*vide infra*). On the other hand, the kinetic energy of the particles which is proportional to their masses, is different from zero if the latter are nonzero. Consequently, if the total energy of the universe is equal to zero, it is necessary that the masses and the other properties which are connected with them (XIV-22) tend to zero from above, the variation rates $\dot{\hbar} / \hbar$, $\dot{a} / a$,... being negative.

Moreover, for physical reasons, ε_0 and μ_0 cannot become infinite or equal to zero ($\varepsilon_0 \mu_0 c^2 = 1$). They must either remain constant or tend to nonzero finite values, so that the charges themselves tend to zero. In other words, the matter which constitutes the universe is vanishing.

On the other hand, the expansion of the universe, assumed as being isotropic and homogeneous, is characterized by the variation rate of its curvature radius, $H = \dot{R} / R$, versus time. The present value of this ratio is called Hubble's constant H_0. Recent determinations show that H_0 is located within the range 50-100 $km\ s^{-1}\ Mpc$, i.e. $5\text{-}10 \times 10^{-11}\ year^{-1}$ [19]. According to (XIV-6), applied to $U = R$, we obtain

$$(H_0)_{obs} = (H_0)_{act} - \frac{\dot{a}}{a} \tag{XIV-31}$$

Consequently, if the universe is effectively expanding, i.e. if $(H_0)_{act} > 0$, then

$$-(H_0)_{obs} < \frac{\dot{a}}{a} \tag{XIV-32}$$

Thus, at the present time,

$$-2 \times 10^{-10} \, yr^{-1} < \left(\dot{\hbar} / \hbar\right)_0 < 0 \qquad \text{(XIV-33)}$$

Let us assume that $\hbar$ behaves as R^{-n}, i.e. as $t^{-2n/3}$ when t tends to infinity. Then

$$\left(\frac{\dot{\hbar}}{\hbar}\right)_0 = -\frac{2n}{3t_0} \qquad \text{(XIV-34)}$$

t_0 being the age of the universe, estimated at $15\text{-}20 \times 10^9$ yrs. From which

$$0 < n < 4 \qquad \text{(XIV-35)}$$

Consequences and various applications

Independently of the fact that, if the fundamental parameters tend to zero, the universe is vanishing, it is interesting to examine some practical consequences of these variations.

As we have seen, even if these parameters vary versus time, all the physical laws are preserved, provided that relationship (XIV-22) is respected. Moreover, the measurement of a property of a body (e.g. : mass, size) leads to the same result whatever the epoch of the measurement may be, because the measurement result is equal to the value of the property itself divided by that of the corresponding unit (XIV-1) . Likewise for the Compton wave-lenght $\hbar / mc$ and the Thomson cross-section $\propto e^4 / m^2 c^4$. And still more, the chemical and the nuclear behaviors of matter are unchanged. The former, indeed, are governed by the ratios W / kT , where W is the corresponding activation energy ($\propto e^2 / \varepsilon_0 a$), proportional to k , which remain constant, and the latter by the values of the ratios of the electromagnetic, strong and weak interaction energies, respectively proportional to $e^2 / \varepsilon_0 a$, g_s^2 / a and $\hbar^3 / Mg_w$ (i.e. to $\hbar^{1/2}$) which are also invariant. Consequently, the chemical evolution of the universe is unaffected. In particular, the temperature at which protons and electrons of the early universe condensed to form neutral atoms, remains unchanged.

Lastly, to answer to the question asked in the title of this chapter, no variation of the Planck constant can be detected because the action unit, namely $m\,a\,c$ (XIV-14) , is precisely proportional to h .

The case of the Stephan constant which increases as the Planck constant decreases, can appear as being disturbing. In fact, the difficulty is not real. The power radiated by a black body, indeed, is proportional to $\sigma a^2 T^4$, a being a characteristic dimension of the latter. Now, according to (XIV-22) , σa^2 is time-invariant. Consequently, if the black body under consideration receives a constant power (the power unit , proportional to mc^2 / τ_0 is time-independent) , its temperature T is unchanged.

A great deal of other similar examples could be given, so that, at the limit, we can be tempted to wonder whether the problem of the past variability of the fundamental parameters of the universe possesses a true physical meaning, because this variability would not modify our knowledge of the word. In fact, the problem is not a pure mathematical play. Astronomy offers interesting application examples.

Owing to the expanding universe, a radiation whose wavelength is equal to λ_e at its emission time, reaches us with the following wavelength $\lambda_{obs} = \lambda_e (R_e / R_0)$, R_0 being the present curvature radius of the universe, and R_e that at the emission time. Let λ_0 be the wavelength we observe on Earth for the corresponding radiation.

Given that λ is proportional to a (XIV-8) , we obtain the following *spectral ratio*

$$\bar{\varsigma} = \frac{\lambda_{obs}}{\lambda_0} = \frac{\lambda_e}{\lambda_0} \cdot \frac{R_0}{R_e} = \frac{(R/a)_0}{(R/a)_e} = \frac{[R]_0}{[R]_e} \tag{XIV-36}$$

while the classical theory gives

$$\varsigma = \frac{R_0}{R_e} \tag{XIV-37}$$

The observed spectral ratios being always superior to 1 (redshift phenomenon), the classical theory concludes that the radius of the universe is increasing versus time $(R_0 > R_e)$. In fact, we can only assert that the ratio R/a , i.e. the value of R in unit a , is increasing versus time. From (XIV-36) and (XIV-37) , it results [12]

$$\tilde{\varsigma} = \varsigma \frac{a_e}{a_0} \qquad\qquad (XIV\text{-}38)$$

Likewise, one would show that the temperature T of the universe varies versus the expansion according to the law

$$T\frac{R}{a} = Cst \qquad\qquad (XIV\text{-}39)$$

while the classical theory gives $T R = Cst$.

Concerning the Hubble constant, strictly speaking, according to (XIV-31) , its actual value must be weaker than the observed one

$$H_{act} = \frac{2}{2+n} H_{obs} \quad \text{i.e.} \quad H_{act} = (2/3 \pm 1/3) H_{obs} \qquad (XIV\text{-}40)$$

if we adopt the law $R \propto t^{-n}$ with $0 < n < 4$. Thus the age of the universe would be greater than that deduced from H_{obs}. In fact, the values obtained for H_{obs} exhibit a notable dispersion $(\pm 30\%)$, so that the introducing of the corrective term $\dot{a}/a$ would be illusory.

The variation of the Earth-Moon distance has been proposed as a test for the variability of the fundamental parameters [12] . The variation rate of h would be of about $-10^{-11} yr^{-1}$. Unfortunately, on the one hand, the present measurements are not sufficiently precise and, on the other, it is difficult to isolate the possible variation due to the variability of the constants themselves from that arising from the expanding universe and from the tide effects, so that no definitive conclusion can be obtained. In any case, a variation of the constants is not excluded.

Remark about the $\hbar \to 0$ limit in quantum mechanics

In order to try to find the classical mechanics again from quantum mechanics, the first idea which occurs to the mind, is to make h formally tending to zero in the equations of quantum mechanics. In fact, such an operation is not licit because h is not a simple parameter exterior to the system, as quantum mechanics assumes explicitly. It is strongly connected with the existence of the matter and more especially with the system under consideration. According to (XIV-22) , we are not entitled to make the Planck constant vary arbitrarily whereas the charge and the mass of electron would be assumed as remaining constant.

References

[1] E. Mach, *Die Geschichte und die Wurzel des Satzes von der Erhaltung der Arbeit* (Prag) 1872 .

[2] A. Eddington, *Space, Time and Gravitation* (Cambridge) 1921.

[3] G. Gamov, *Phys. Rev. Lett.* **19** (1967) 759.

[4] P.A.M. Dirac, *Nature* **139** (1937) 323 ; *Proc. Roy. Soc. London* **A 165** (1938) 199.

[5] V. Canuto, J. Lodenquai, *Appl. J.* **211** (1977) 342.

[6] T.C. van Flandern, *Month. Not. Roy. Ast. Soc.* **170** (1953) 333.

[7] R.D. Reasenberg, I.I. Shapiro, *Bound on the secular variation of the gravitational interaction* (MIT preprint) 1975.

[8] E.J. Dyson, in *Current Trends in the Theory of Fields* , Proceedings of AIP Conference n°48, Edited by J.E. Lannutti and P.K. Williams (AIP, New York) 1978, p. 163.

[9] B.E.J. Pagel, *Month. Not. Roy. Ast. Soc.* **179** (1977) 81.

[10] J.P. Petit, *Modern Phys. Lett.* **A 16** (1989) 1527, 1733.

[11] A. Julg, *Annales Fond. Louis de Broglie* **14** (1989) 301.

[12] A. Julg, *Hadronic Journal Suppl.* **8** (1993) 415.

[13] W.A. Baum, R. Florentin-Nielsen, *Ast. J.* **209** (1976) 319.

[14] P.D. Noerdlingen, *Phys. Rev. Lett.* **30** (1973) 761 ; D.P. Woody , P.L. Richards, *Phys. Rev. Lett.* **42** (1979) 925.

[15] E. Cartan, *J. Math. Pures et Appliquées* **1** (1922) 141.

[16] L.H. Ryder, *Quantum Field Theory* (Cambridge Univ. Press, Cambridge) 1984, p. 241 (Caution! The formula is given in units $\hbar = c = 1$).

[17] C.W. Misner, K.S. Thorne, J.A. Wheeler, *Gravitation* (W.H. Freeman and Co, New York) 1973.

[18] For instance, see: P.C.W. Davies, *The accidental universe* (Cambridge Univ. Press, Cambridge) 1982.

[19] A. Sandage, in *Proceedings of the Symposium on the Galaxy and the Distance Scale* (Essex, England) 1972 ; G. de Vaucouleurs, *Colloques CNRS* , n°37 (Paris) 1976 ; N. Bartel, A.E.E. Rogers, I.I. Shapiro, M.V. Gorenstein, C.R. Gwinn, J. M. Marcaide, K.W. Weiler, *Nature* **318** (1985) 25 ; A.P. Marsher, *Nature* **318** (1985) 18.

Conclusion

At the end of this report, the chief conclusion which emerges from it, is that, contrary to common opinion, so as not to say contrary to the dogma which is taught, the microscopic phenomena can be interpreted by means of the same concepts as those utilized to describe the phenomena which occur at our scale. Physics finds its conceptual unity again because the same laws govern the world of atoms as well as that of galaxies.

Starting from the idea that any system is never strictly isolated, but is unceasingly exchanging energy with the other systems which constitute the universe, we have built up a model essentially classical in its structure, which, not only is able to account for experiment, but also can be transcribed into an operator formalism, analogous to that *a priori* put forth by quantum mechanics. At first sight, our model would seem to constitute a justification of the latter. In fact, in our model, and consequently in the formalism which arises from it, no dynamical property exhibits a constant value, but it is unceasingly fluctuating versus time around a certain value which is precisely that given by the operator formalism, while in quantum mechanics certain properties, energy in particular, remain constant versus time. Strictly speaking, our model cannot be utilized as a basis for interpreting the quantum mechanics, at least in its *orthodox* form.

This restriction is very important, because, if, to avoid a certain number of mathematical and logical difficulties which are inherent in the conventional quantum formalism, we symmetrize the classical expressions of the various dynamical properties with respect to the canonical variables q and p before constructing the corresponding operators, all the properties, without exception, do vary around the values given by these operators. Thus, our model corresponds to this new version of quantum mechanics. The difference between the latter and that which our model suggests to us, is, nevertheless, sufficiently weak for us to be entitled to consider our symmetrized formalism as being still quantum-like.

But a formalism is neither more nor less than a tool, which, because of this, cannot claim to correspond to reality and still less to give a direct image of the latter. Thus, although the quantum formalism forbids us to approach the fine behavior of any system (e.g. : we cannot follow the motion of the particle along its trajectory as in classical mechanics, i.e. we cannot know its position and its speed at any time) , we are not entitled to ascribe the lack of information about the behavior of the system to any physical cause which would bound our knowledge and, still less, to decree that what not appears in quantum mechanics (for instance, the trajectory notion) has no physical meaning.

Our model which leads to a formalism very close to that of the orthodox quantum mechanics, clearly shows this point. As does the orthodox formalism, the symmetrized one does not allow to reach a certain number of properties of the system, the motion on the trajectories in particular. But we see clearly that the information loss is due to the mathematical reduction of the problem when we pass to operators, and not to any physical necessity. It is the impossibility to work on the trajectories Γ and Z themselves, which constrains us to have recourse to an operator-like treatment which allows to tackle the system as a whole without needing to know its detailed evolution.

But this reduction is possible only because the physical phenomena exhibit a quasi-ergodic nature arising from the permanent reciprocal interaction between the systems which constitute the universe. We have developed our model for microsystems, but it is obvious that the quasi-ergodic character is not just an attribute of systems of small size. Every system, whatever its size exhibits this character. Nevertheless, the fluctuations of which the system is the seat, play a role at the microscopic level only, so that every thing occurs for us as if these fluctuations did not exist. That is, finally, very fortunate, because it was possible to establish very simple basic laws of physics precisely by studying the behavior of macroscopic bodies for which fluctuations are not observable, before coming up against problems when one has tried to apply these laws, as they are, at the microscopic level.

In any case, it is important to realize that these fluctuations have no common points with the fundamental indeterminism ascribed by the orthodox quantum interpretation to the behavior of any system. They arise from the strict application of completely deterministic equations *at the universe scale* . From which results a certain unforeseeable character of the phenomena which is due to the practical impossibility to tackle the universe as a whole.

Besides, the so-called quantum indeterminism arises, in fact, from a more philosophical than physical abusive interpretation of a purely mathematical relationship, namely the Heisenberg one, whose very meaning is completely different from that which one took pleasure in giving to it. To claim that, according to this relationship, the greater the accuracy obtained for the position, the weaker that for the speed, not only is unjustified, but also it is at variance with the quantum formalism from which this relationship is issued, which is still worse! This relationship, indeed, says only that the product $(\Delta x)^2 (\Delta p_x)^2$ of the average quadratic dispersions for the position and the momentum respectively, relating to any system, is greater than or equal to $\hbar^2 / 4$. Now, for a given system, $(\Delta x)^2$ and $(\Delta p_x)^2$ exhibit well-determined values, for instance, $\hbar / 2m\omega$ and $\hbar m\omega / 2$ in the harmonic oscillator for which we have the equality. These two terms cannot be interpreted as uncertainties which, by definition, would be able to exhibit arbitrary values.

In any case, the most important conclusion about the understanding of the physical world is certainly the fundamental role played by the interactions between the systems which constitute the universe. If this role does not directly appear in the quantum formalism, it is, nevertheless, present in the equations through the Planck constant, quantum mechanics being only a way of reducing the too complex problem of a system in interaction with an extremely great number of other systems to the problem of the system itself by means of a mathematical artifact. In the absence of these interactions, no system would be stable, so that the existence of the least fragment of organized matter, if only an atom or a living cell, at a given point of the universe results from the presence of the other systems which constitute the universe. This outlook is broader than that of Mach [1] for which, schematically speaking, *in the universe, every thing acts on the others* , because our interpretation involves that these interactions are precisely the very origin of the organization of the universe, such as we know it.

But such an interdependence between matter and its own existence is not limited to complex systems (molecules, atoms, nucleons or hadrons). It appears also at the level of the fundamental particles (leptons and quarks). We have, indeed, seen that the relationships concerning the parameters of physics and their behavior in relation to the expansion of the universe, involve that the mass and the charge of particles tend to zero as the universe radius increases. This result is the direct consequence of this interdependence. The old idea of Mach according to which the mass of particles could arise from their mutual interaction [1] , would here find a justification.

Destined to vanish, matter would not be eternal, which states the problem of its nature with still more acuteness.

Reference

[1] E. Mach, *Die Geschichte und die Wurzel des Satzes von der Erhaltung der Arbeit* (Prag) 1872, Engl. Transl.: *History and root of the principle of conservation of energy* (Open Court, Chicago) 1911 ; *The Science of mechanics* (Open Court, La Salle) 1960.

Appendix :
Coupled systems and ergodicity

An example of coupled system

Let us consider two pendulums (1 and 2) connected by a weak spring, or attached to a common support in such a way that interaction of the two occurs by way of the support. The equations which govern the motion of the two pendulums are

$$\begin{cases} \ddot{x}_1 + \omega_1^2 x_1 = k(x_1 - x_2) \\ \ddot{x}_2 + \omega_2^2 x_2 = k(x_2 - x_1) \end{cases} \tag{A-1}$$

ω_1 and ω_2 being the respective eigenfrequencies of the pendulums. One says that these oscillators form a *coupled system* , k being the coupling constant.

By eliminating x_2 between the two equations (A-1) we obtain

$$\ddot{x}_1 + (\omega_1^2 + \omega_2^2 - 2k)\ddot{x}_1 + \left[\omega_1^2\omega_2^2 - k(\omega_1^2 + \omega_2^2)\right]x_1 = 0 \tag{A-2}$$

Let us put

$$x_1 = e^{\lambda t} \tag{A-3}$$

We have

$$\lambda^4 + (\omega_1^2 + \omega_2^2 - 2k)\lambda^2 + \left[\omega_1^2\omega_2^2 - k(\omega_1^2 + \omega_2^2)\right] = 0 \tag{A-4}$$

Let $\lambda_1^2 = -\alpha^2$ and $\lambda_2^2 = -\beta^2$ be the roots of this equation. The general solution of (A-2) is

$$x_1 = A\sin\alpha t + B\cos\alpha t + C\sin\beta t + D\cos\beta t \tag{A-5}$$

A , B , C and D being constants determined by the initial conditions.

For instance, let us assume that at the time $t = 0$, we have

$$(x_1)_0 = a, \ (\dot{x}_1)_0 = 0, \ (x_2) = 0 \text{ and } (\dot{x}_2) = 0 \tag{A-6}$$

Then we obtain

$$A = C = 0 \ , \ B = \frac{\omega_1{}^2 - \beta^2 - k}{\alpha^2 - \beta^2} a \ , \ D = -\frac{\omega_1{}^2 - \alpha^2 - k}{\alpha^2 - \beta^2} a \qquad \text{(A-7)}$$

Consequently, the motion of pendulum 1 is the superimposition of two harmonic motions of respective periods $2\pi/\alpha$ and $2\pi/\beta$

$$x_1 = B\cos\alpha t + D\cos\beta t \qquad \text{(A-8)}$$

The motion is *periodical* only if the ratio of the two periods is equal to $p\ /q$, p and q being integers.

In the opposite case, the motion exhibits an irregular behavior which will appear to us as being completely *chaotic* when the number of coupled systems becomes sufficiently great. For instance, for a linear set of N oscillators

$$x_1 = \sum_{k=1}^{N} A_k \sin(\alpha_k t - \varphi_k) \qquad \text{(A-9)}$$

Ergodic character of coupled systems

Let us consider a system such as the time-average values of the various properties $G\ (t\ )$ attached to the system tend versus time to finite limits $\overline{G}$, i.e. that for each property we have

$$\left| \frac{1}{\tau} \int_0^\tau G(t)dt - \overline{G} \right| < \varepsilon \qquad \text{(A-10)}$$

when $\tau > \tau_e$, ε being an arbitrarily small quantity.

One says that the system exhibits an *ergodic* or rather *quasi-ergodic* behavior. The reader will find the detailed theory of such systems in Ref [1] .

The coupled systems fall under this category as it is easy to verify in the case of the two oscillators we have considered. For instance, according to (A-8) , the time average value of the kinetic energy $T_1(t) = \frac{1}{2} m_1 \dot{x}_1{}^2$ (m_1 being the mass of the oscillator) is the following

$$\frac{1}{\tau}\int_0^\tau \frac{m_1}{2}(\alpha B \sin \alpha t + \beta B \sin \beta t)^2 \, dt = \frac{m_1}{4}\left(\alpha^2 B^2 + \beta^2 D^2\right)$$

$$-\frac{m_1}{4\tau}\left\{\alpha B^2 \sin 2\alpha\tau + \beta D^2 \sin 2\beta\tau + 2\alpha\beta BD\left(\frac{\sin(\alpha+\beta)\tau}{\alpha+\beta} - \frac{\sin(\alpha-\beta)\tau}{\alpha-\beta}\right)\right\}$$

(A-11)

which tends to $\frac{m_1}{4}\left(\alpha^2 B^2 + \beta^2 D^2\right)$ when τ tends to infinity, with

fluctuations whose amplitude decreases in the mean as $1/\tau$.

Likewise for any property depending on x_1, e.g. the potential energy $U_1 = \frac{1}{2}m_1\omega_1^2 x_1^2$.

Such a behavior occurs, of course, in harmonic oscillator. Nevertheless in the latter the total energy (kinetic + potential) remains constant. In the case of coupled systems a given oscillator is not isolated so that its energy fluctuates versus time. For instance, for pendulum 1, we have

$$E_1 = \frac{1}{2}m_1\dot{x}_1^2 + \frac{1}{2}m_1\omega_1^2 x_1^2 \tag{A-12}$$

which tends to $\frac{m_1}{4}\left[\left(\alpha^2 B^2 + \beta^2 D^2\right) + \omega_1^2\left(B^2 + D^2\right)\right]$ when τ tends to

infinity, with fluctuations whose amplitude tends to zero.

Probability density

Amongst all the properties of the systems which exhibit an ergodic behavior, the most interesting for our problem is certainly the existence of a *probability density* which allows to replace the time-average value of a property G attached to the system by a space-mean.

In order to proof this proposition, we will consider an 1D-system constituted by one particle.

Let $x(t)$ be the position of the particle at the time t ($x \in D$). Let us perform the mean of property G over the time interval $(0,\tau)$. Given that the particle passes a great number of times at a given position x , the inversion of the function x (t) which would give $t = t(x)$ is impossible. In order to turn the difficulty, we will cut the trajectory into segments k on which the variation of x versus t is monotonic. At each position x of the segment k we can define a presence probability $\pi_k(x)$. By regrouping all the segments to reconstruct the complete trajectory, for each position x we obtain the probability density

$$\sum_k \pi_k(x) = \rho_\tau(x) \tag{A-13}$$

Then, the average value of the property G depending on x over the time interval $(0,\tau)$ is $\left(\overline{G}\right)_\tau = \frac{1}{\tau} \int_0^\tau G[x(t)]\, dt = \int_D G(x)\rho_\tau(x)dx$ (A-14)

Now, by hypothesis, when τ tends to infinity, $\left(\overline{G}\right)_\tau$ tends to $\overline{G}$. Consequently, $\rho_\tau(x)$ tends to a finite limit $\rho(x)$ which is the *probability density* to find the particle at the point x , so that we can write

$$\overline{G} = \int_D G(x)\rho(x)dx \tag{A-15}$$

Likewise, we can define a density $\rho'(\mathbf{p})$ in the space of the momenta $\{\mathbf{p} = m\mathbf{v}\}$ for properties depending on the momenta only.

As a direct example, we will give the proof of these propositions in a very simple case, namely that of a saw-tooth function defined by

$$x_k = v(t - kT) \quad \text{over } [kT,(k+1)T] \quad (k = 0,1,2...) \tag{A-16}$$

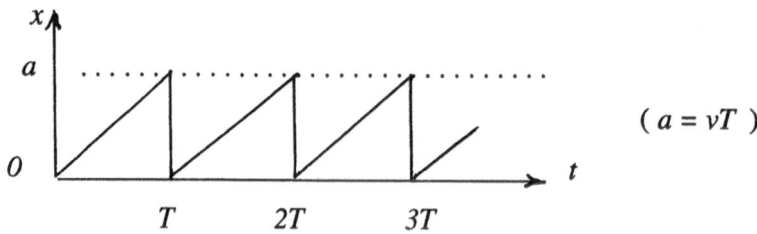

$(a = vT)$

In this view, let us perform the average value of x^2 over the time interval $(0,\tau)$, assumed to be sufficiently large for being considered as practically equal to an integer of periods T : $\tau = n\,T$.

$$\overline{x^2} = \frac{1}{nT}\sum_k \int_{kT}^{(k+1)T} v^2t^2\,dt = \frac{v^2}{nT}\times n \times \frac{T^3}{3} = \frac{v^2T^2}{3} = \frac{a^2}{3} \qquad \text{(A-17)}$$

On the other hand, the time passed by the particle over the segment $(x - dx/2 \,, x + dx/2)$ is equal to $dx \ /v$. Over the interval $\tau = nT$, the total time passed by the particle on the segment under consideration is

$$n\frac{dx}{v} = nT\frac{dx}{a} \qquad \text{(A-18)}$$

Consequently, the probability to find the particle on the segment under consideration is

$$\frac{1}{nT}\times\left(n\frac{dx}{v}\right) = \frac{dx}{a} \qquad \text{(A-19)}$$

On the other hand, the vertical segments of the curve x $(t$) bring no contribution owing to the fact that on the latter the velocity is infinite.

Hence, we obtain the following probability density

$$\rho(x) = \frac{1}{a} \qquad \text{(A-20)}$$

and the average value of the squared position

$$\overline{x^2} = \int_0^a x^2\rho(x)\,dx = \frac{a^2}{3} \qquad \text{(A-21)}$$

which, as expected, is equal to the value obtained in (A-17) .

Reference

[1] A. Blanc-Lapierre, P. Casal, A. Tortrat, *Méthodes mathématiques de la Mécanique Statistique* (Masson, Paris) 1959.

INDEX

Lecture Notes in Chemistry

For information about Vols. 1–29
please contact your bookseller or Springer-Verlag

Editorial Policy

This series aims to report new developments in chemical research and teaching - quickly, informally and at a high level. The type of material considered for publication includes:

1. Preliminary drafts of original papers and monographs

2. Lectures on a new field, or presenting a new angle on a classical field

3. Seminar work-outs

4. Reports of meetings, provided they are
   a) of exceptional interest and
   b) devoted to a single topic.

Texts which are out of print but still in demand may also be considered if they fall within these categories.

The timeliness of a manuscript is more important than its form, which may be unfinished or tentative. Thus, in some instances, proofs may be merely outlined and results presented which have been or will later be published elsewhere. If possible, a subject index should be included. Publication of Lecture Notes is intended as a service to the international chemical community, in that a commercial publisher, Springer-Verlag, can offer a wider distribution to documents which would otherwise have a restricted readership. Once published and copyrighted, they can be documented in the scientific literature.

Manuscripts

Manuscripts should comprise not less than 100 and preferably not more than 500 pages. They are reproduced by a photographic process and therefore must be submitted in camera-ready form according to Springer-Verlag's specifications: technical instructions will be sent on request.

The text area should take care of the page length and width (12.2 x 19.3 cm when you use a 10 point font size, 15.3 x 24.2 cm for a 12 point font size).

Authors receive 50 free copies and are free to use the material in other publications.

Manuscripts should be sent to one of the editors or directly to Springer-Verlag, Heidelberg.

Springer
and the
environment

At Springer we firmly believe that an international science publisher has a special obligation to the environment, and our corporate policies consistently reflect this conviction.
We also expect our business partners – paper mills, printers, packaging manufacturers, etc. – to commit themselves to using materials and production processes that do not harm the environment. The paper in this book is made from low- or no-chlorine pulp and is acid free, in conformance with international standards for paper permanency.

 Springer